Forum for Interdisciplinary Mathematics

The *Forum for Interdisciplinary Mathematics* is a Scopus-indexed book series. It publishes high-quality textbooks, monographs, contributed volumes and lecture notes in mathematics and interdisciplinary areas where mathematics plays a fundamental role, such as statistics, operations research, computer science, financial mathematics, industrial mathematics, and bio-mathematics. It reflects the increasing demand of researchers working at the interface between mathematics and other scientific disciplines.

Xiang-Sheng Wang • Chisheng Wang

Machine Learning in Data Processing

Springer

Xiang-Sheng Wang
University of Louisiana at Lafayette
Lafayette, LA, USA

Chisheng Wang
Shenzhen University
Shenzhen, Guangdong, China

ISSN 2364-6748 ISSN 2364-6756 (electronic)
Forum for Interdisciplinary Mathematics
ISBN 978-3-032-20854-5 ISBN 978-3-032-20855-2 (eBook)
https://doi.org/10.1007/978-3-032-20855-2

Mathematics Subject Classification: 62J07, 15A18, 65K05, 68T07, 90C25

This Springer imprint is published by the registered company Springer Nature Switzerland AG
The registered company address is: Gewerbestrasse 11, 6330 Cham, Switzerland

To Dongying Xu

Preface

Machine learning is perceived in diverse ways across academic and professional communities, with each perspective influenced by disciplinary background and practical experience. Rooted in mathematics, statistics, computer science, data science, and numerical computation, the field has advanced rapidly in the era of big data. Its broad impact on science, engineering, and everyday life has led many universities to actively integrate machine learning into both undergraduate and graduate curricula.

At the request of several graduate students, the first author began teaching a course on machine learning in Fall 2022 at the University of Louisiana at Lafayette. Offered through the Department of Mathematics, the course attracted students from a broad range of disciplines, including mechanical engineering, systems engineering, electrical engineering, earth and energy sciences, and industrial chemistry. Many of these students had substantial hands-on experience applying machine learning techniques in data processing but wished to gain a deeper understanding of the underlying mathematical principles that explain why these methods work so effectively.

In contrast, most mathematics majors enrolled in the course had little or no prior exposure to machine learning or computer programming and often lacked experience writing even simple code. Their interest was motivated by the growing prominence of machine learning and a desire to understand foundational concepts such as neural networks and support vector machines from a mathematical standpoint.

To accommodate this diverse student population, the first author consulted with the second author, a geo-informatics researcher specializing in the application of mathematical and data analysis techniques to geoscience problems. Together, the authors reviewed numerous textbooks and online resources. While many of these materials are well suited for engineering students focused on applications, they are often too advanced for mathematics students with limited programming background. Moreover, existing resources tend to emphasize large real-world datasets and extensive coding, often at the expense of clear and systematic mathematical exposition.

This lack of suitable instructional material motivated the authors to write a textbook specifically designed for students with minimal programming experience. The goal of the book is to provide a mathematically grounded introduction to machine learning without requiring extensive coding skills.

During the 2023–2024 academic year, the first author took a sabbatical and collaborated closely with the second author to substantially revise and expand the lecture notes from the Fall 2022 course. This collaboration resulted in the completion of the initial draft of the book. Upon returning from sabbatical, the first author used the manuscript as the primary text for a redesigned machine learning course in Fall 2024. The chapters were structured to fit a standard one-semester course of 30–40 contact hours, and an appendix was added to review essential topics in calculus and linear algebra.

This book consolidates fundamental mathematical derivations and theoretical results that are otherwise scattered throughout the machine learning literature. Unlike many existing texts that focus on large-scale real-world datasets, this book emphasizes simplified synthetic data to clearly illustrate core concepts and methods. While coding is optional, concise Python scripts are provided to allow students to experiment with examples and modify them for exercises.

The only prerequisites for this book are a solid understanding of calculus and linear algebra. Key topics such as Taylor expansions, derivatives, the chain rule, and matrix operations are used extensively. For example, the derivation of backward propagation in neural networks relies heavily on the chain rule and matrix calculus. Basic statistical concepts—such as mean, variance, probability distributions, and probability density functions—are occasionally referenced, and familiarity with elementary statistics will aid comprehension.

This book focuses primarily on supervised learning, where models are trained using labeled data, consisting of input–output pairs, and their performance is evaluated by their ability to generalize from training data to previously unseen test data. While this book does not treat them in detail, we emphasize that other major paradigms of machine learning also exist, including unsupervised learning, which seeks structure in unlabeled data, and reinforcement learning, which studies decision-making through interaction with an environment.

Throughout the book, learning is formulated as an optimization problem, and model performance must be assessed in terms of prediction accuracy and generalization error. Issues such as overfitting are discussed from a mathematical perspective, motivating regularization techniques and model selection. The text is intentionally selective: it concentrates on regression models, classification methods, support vector machines, and neural networks, while omitting topics such as probabilistic graphical models and reinforcement learning in order to maintain a coherent and rigorous mathematical narrative. The goal is not to provide a comprehensive survey of all machine learning methods, but to equip readers—particularly those with strong mathematical or computational background—with a clear understanding of the numerical methods and theoretical principles that underlie modern supervised learning.

This book is intended for two primary audiences: students majoring in mathematics or statistics who are not actively working in machine learning but wish to understand its fundamental concepts and terminology, and students from other disciplines who have practical experience using machine learning tools and seek a rigorous understanding of the mathematical foundations underlying these techniques.

A central objective of this book is to promote machine learning as a standard undergraduate or graduate course—comparable to differential equations—requiring minimal prerequisites in programming and statistics. By relying primarily on calculus and linear algebra, the material is accessible to students who have completed basic coursework in these subjects.

To achieve this accessibility, the presentation is deliberately elementary and self-contained. Upon completing this book or a course based on it, students should acquire foundational knowledge of linear and nonlinear regression, regularization, neural networks, batch normalization, support vector machines, gradient-based optimization methods, and principal component analysis. The exposition follows a theorem-and-proof style to ensure mathematical rigor and clarity.

Python is chosen as the programming language because it is free, widely available, and easy to read. Although programming languages evolve over time, Python's simplicity makes it particularly suitable for beginners. Code examples are intentionally kept short—typically no more than 30–40 lines—and avoid reliance on specialized libraries.

The first author gratefully acknowledges the unwavering support of his wife, Hongying, and his children, Mingqian and Tongwei. The second author gratefully acknowledges the constant encouragement of his wife, Yanhong, and his children, Yuchen and Xihe. The authors are deeply thankful for their families' support, which enabled them to persevere through the challenges of completing this book.

Lafayette, LA, USA — Xiang-Sheng Wang
Shenzhen, China — Chisheng Wang

Contents

1 Matrix 1
- 1.1 Notations 1
- 1.2 Determinant 3
- 1.3 Block Matrix 6
- 1.4 Rank 9
- 1.5 Fréchet Derivative 12
- Problems 14

2 Linear Regression 17
- 2.1 Notations 17
- 2.2 Linear Model 18
- 2.3 Maximum Likelihood Estimation 19
- 2.4 Least Squares Approximation 20
- 2.5 Sum of Squared Errors 21
- 2.6 Variance Inflation Factor 23
- 2.7 Python Code 25
- Problems 27

3 Regularization 31
- 3.1 Overfitting Problem 31
- 3.2 Ridge Regression 33
- 3.3 Convex Optimization 34
- 3.4 LASSO 38
- 3.5 Python Code 39
- 3.6 Discussions 40
- Problems 41

4 Nonlinear Regression 45
- 4.1 Nonlinear Data 45
- 4.2 Sigmoid Function 46
- 4.3 Optimization Problem 48
- 4.4 Gradient Descent Method 49

4.5 Python Code ... 50
4.6 Discussions ... 51
Problems ... 52

5 Shallow Neural Network ... 55
5.1 Notations ... 55
5.2 The XOR Gate ... 56
5.3 Composite Function ... 57
5.4 Python Code ... 59
5.5 Discussions ... 59
Problems ... 60

6 Deep Neural Network ... 63
6.1 The Model ... 63
6.2 Backward Propagation ... 64
6.3 Python Code ... 65
6.4 Discussions ... 66
Problems ... 68

7 Batch Normalization ... 69
7.1 The Model ... 69
7.2 Backward Propagation ... 70
7.3 Python Code ... 72
7.4 Discussions ... 73
Problems ... 74

8 Support Vector Machine ... 75
8.1 Notations ... 75
8.2 Optimization Problems ... 76
8.3 Equivalence Theorem ... 79
Problems ... 80

9 Gradient Methods ... 83
9.1 Conjugate Gradient Method ... 83
9.2 One-Step Method ... 86
9.3 Two-Step Method ... 88
Problems ... 90

10 Dimensionality Reduction ... 93
10.1 Schur Decomposition ... 93
10.2 Singular Value Decomposition ... 94
10.3 Principal Component Analysis ... 95
Problems ... 100

A Related Topics ... 103
A.1 Calculus ... 103
A.2 Linear Algebra ... 106

B Hints to Selected Exercise Problems 111

C Further Readings 113

References 115

Index 117

Chapter 1
Matrix

Abstract In this chapter, we review some terminologies and properties of matrices which will be used throughout this book. For simplicity, we mainly focus on real matrices, even though many statements and proofs can be easily extended to matrices in complex number system.

1.1 Notations

Throughout this chapter, we denote by $\mathbb{R}^n$ the space of n-dimensional real vectors. By default, a vector $v \in \mathbb{R}^n$ is a *column vector* with components $v_1, \cdots, v_n$. A collection of m vectors in $\mathbb{R}^n$ forms a *matrix* $A = (A_1, \cdots, A_m)$ of dimension $n \times m$, where

$$A_k = \begin{pmatrix} A_{1k} \\ \vdots \\ A_{nk} \end{pmatrix} \in \mathbb{R}^n, \ \ k = 1, \cdots, m.$$

The space of all $n \times m$ matrices is denoted by $\mathbb{R}^n \times \cdots \times \mathbb{R}^n = \mathbb{R}^{n\times m}$ (m copies).

Given $A \in \mathbb{R}^{n\times m}$ with $n > 1$ and $m > 1$, the *(i,j)-minor* of A is the submatrix $A^{(i,j)} \in \mathbb{R}^{(n-1)\times(m-1)}$ obtained by deleting the i-th row and j-th column of A. Explicitly,

$$A_{kl}^{(i,j)} = A_{k'l'}, \ \ 1 \le k \le n-1, \ 1 \le l \le m-1,$$

where

$$k' = \begin{cases} k, & k < i, \\ k+1, & k \ge i, \end{cases} \quad l' = \begin{cases} l, & l < j, \\ l+1, & l \ge j. \end{cases}$$

X.-S. Wang, C. Wang, *Machine Learning in Data Processing*, Forum for Interdisciplinary Mathematics, https://doi.org/10.1007/978-3-032-20855-2_1

The *transpose* of a matrix $A \in \mathbb{R}^{n\times m}$ is the matrix $A^T \in \mathbb{R}^{m\times n}$ defined by

$$A^T_{jk} = A_{kj}, \quad 1 \le j \le m, \ 1 \le k \le n.$$

In particular, a column vector $x \in \mathbb{R}^n = \mathbb{R}^{n\times 1}$ can be viewed as an $n \times 1$ matrix and its transpose $x^T \in \mathbb{R}^{1\times n}$ is a row vector. A square matrix $A \in \mathbb{R}^{n\times n}$ is *symmetric* if $A^T = A$; namely, $A_{jk} = A_{kj}$ for any $1 \le j, k \le n$. The *trace* of a square matrix $A \in \mathbb{R}^{n\times n}$ is defined as

$$\mathrm{tr}(A) = \sum_{k=1}^{n} A_{kk}.$$

The *product* of two matrices $A \in \mathbb{R}^{n\times m}$ and $B \in \mathbb{R}^{m\times p}$ is the matrix $AB \in \mathbb{R}^{n\times p}$ given by

$$(AB)_{ij} = \sum_{k=1}^{m} A_{ik}B_{kj}, \quad 1 \le i \le n, \ 1 \le j \le p.$$

The *standard basis* of $\mathbb{R}^n$ consists of vectors $e_1, \cdots, e_n$, where the j-th component of e_k is

$$\delta_{jk} = \begin{cases} 1, & j = k, \\ 0, & j \ne k. \end{cases}$$

The *identity matrix* is $I = (e_1, \cdots, e_n) \in \mathbb{R}^{n\times n}$. It is easily seen that $AI = IA = A$ for any $A \in \mathbb{R}^{n\times n}$. A square matrix $A \in \mathbb{R}^{n\times m}$ is *invertible* (or *nonsingular*) if there exists $B = A^{-1} \in \mathbb{R}^{n\times m}$ such that

$$AB = BA = I.$$

Let $(i_1, \cdots, i_n)$ be any permutation of the indices $(1, \cdots, n)$. Denote by $\sigma(i_1, \cdots, i_n)$ the minimum number of adjacent transpositions required to convert $(i_1, \cdots, i_n)$ into $(1, \cdots, n)$. The *sign* of the permutation is defined as

$$s(i_1, \cdots, i_n) = (-1)^{\sigma(i_1,\cdots,i_n)}.$$

Clearly, the sign function is *antisymmetric*, meaning that

$$s(i_1, \cdots, i_j, \cdots, i_k, \cdots, i_n) = -s(i_1, \cdots, i_k, \cdots, i_j, \cdots, i_n).$$

1.2 Determinant

Given a function f on $\mathbb{R}^{n\times n}$, we say that f is *multilinear* if, for any $k = 1, \cdots, n$, any scalars $\alpha, \beta \in \mathbb{R}$, and any vectors $A_1, \cdots, A_n,\ B_1, \cdots, B_n \in \mathbb{R}^n$,

$$\begin{aligned} &f(A_1, \cdots, A_{k-1}, \alpha A_k + \beta B_k, A_{k+1}, \cdots, A_n) \\ =&\alpha f(A_1, \cdots, A_{k-1}, A_k, A_{k+1}, \cdots, A_n) \\ &+ \beta f(A_1, \cdots, A_{k-1}, A_k, A_{k+1}, \cdots, A_n). \end{aligned} \tag{1.1}$$

We say that f is *antisymmetric* if

$$f(A_1, \cdots, A_j, \cdots, A_k, \cdots, A_n) = -f(A_1, \cdots, A_k, \cdots, A_j, \cdots, A_n), \tag{1.2}$$

that is, f changes sign whenever two of its vector arguments are interchanged. We say that f is *normalized* if

$$f(I) = f(e_1, \cdots, e_n) = 1, \tag{1.3}$$

where $e_1, \cdots, e_n$ are the standard basis vectors of $\mathbb{R}^n$ and $I = (e_1, \cdots, e_n)$ is the identity matrix in $\mathbb{R}^{n\times n}$. The *determinant* of a square matrix $A \in \mathbb{R}^{n\times n}$ is the unique function on $\mathbb{R}^{n\times n}$ that is normalized, antisymmetric, and multilinear. We will show that such a function exists and is unique.

To establish existence, we define f recursively as follows: $f(a) = a$ for $a \in \mathbb{R}$ and

$$f(A) = \begin{cases} A, & A \in \mathbb{R}, \\ \sum_{i=1}^{n} (-1)^{i-1} A_{i1} f(A^{(i,1)}), & A \in \mathbb{R}^{n\times n},\ n > 1. \end{cases} \tag{1.4}$$

When $n = 2$ and $A \in \mathbb{R}^{2\times 2}$, the formula (1.4) yields

$$f(A) = A_{11}A_{22} - A_{21}A_{12}.$$

It is straightforward to verify that f defined in this way is normalized, antisymmetric, and multilinear on $\mathbb{R}^{2\times 2}$.

Lemma 1.1 *The function f defined recursively by* (1.4) *is normalized, antisymmetric, and multilinear on $\mathbb{R}^{n\times n}$.*

Proof We proceed by induction on n. For $n = 1$ and $n = 2$, it is evident that f is normalized, antisymmetric, and multilinear.

Assume that f is normalized, antisymmetric, and multilinear on $\mathbb{R}^{(n-1)\times(n-1)}$. We will show that these properties hold on $\mathbb{R}^{n\times n}$. Let $k = 2, \cdots, n$, $\alpha, \beta \in \mathbb{R}$, and $A_1, \cdots, A_n,\ B_1, \cdots, B_n \in \mathbb{R}^n$. It follows from the recursive definition (1.4) that

$$\begin{aligned}&f(A_1,\cdots,A_{k-1},\alpha A_k+\beta B_k,A_{k+1},\cdots,A_n)\\&=\sum_{i=1}^{n}(-1)^{i-1}A_{i1}f(A_2^{(i)},\cdots,A_{k-1}^{(i)},\alpha A_k^{(i)}+\beta B_k^{(i)},A_{k+1}^{(i)},\cdots,A_n^{(i)}),\end{aligned}$$

where $A_j^{(i)}\in\mathbb{R}^{n-1}$ (respectively, $B_j^{(i)}\in\mathbb{R}^{n-1}$) denotes the vector obtained from A_j (respectively, B_j) by deleting its i-th component. Since f is multilinear on $\mathbb{R}^{(n-1)\times(n-1)}$, we have

$$\begin{aligned}&f(A_1,\cdots,A_{k-1},\alpha A_k+\beta B_k,A_{k+1},\cdots,A_n)\\&=\alpha\sum_{i=1}^{n}(-1)^{i-1}A_{i1}f(A_2^{(i)},\cdots,A_{k-1}^{(i)},A_k^{(i)},A_{k+1}^{(i)},\cdots,A_n^{(i)})\\&\quad+\beta\sum_{i=1}^{n}(-1)^{i-1}A_{i1}f(A_2^{(i)},\cdots,A_{k-1}^{(i)},B_k^{(i)},A_{k+1}^{(i)},\cdots,A_n^{(i)})\\&=\alpha f(A_1,\cdots,A_{k-1},A_k,A_{k+1},\cdots,A_n)\\&\quad+\beta f(A_1,\cdots,A_{k-1},B_k,A_{k+1},\cdots,A_n).\end{aligned}$$

Next, we obtain

$$\begin{aligned}&f(\alpha A_1+\beta B_1,A_2,\cdots,A_n)\\&=\sum_{i=1}^{n}(-1)^{i-1}(\alpha A_{i1}+\beta B_{i1})f(A_2^{(i)},\cdots,A_n^{(i)})\\&=\alpha f(A_1,A_2,\cdots,A_n)+\beta f(B_1,A_2,\cdots,A_n).\end{aligned}$$

Coupling the above two results shows that f is multilinear on $\mathbb{R}^{n\times n}$.

Recall that $e_1,\cdots,e_n$ are the standard basis vectors of $\mathbb{R}^n$. It is easily seen that $e_2^{(1)},\cdots,e_n^{(1)}$ form the standard basis vectors of $\mathbb{R}^{n-1}$, where $e_j^{(1)}$ is the vector obtained by deleting the first component of e_j for $j=2,\cdots,n$. Consequently, we have

$$f(e_1,\cdots,e_n)=f(e_2^{(1)},\cdots,e_n^{(1)})=1,$$

which proves that f is normalized on $\mathbb{R}^{n\times n}$.

Finally, we assume $A_j=A_k$ for some $1\le j<k\le n$. If $j>1$, then the antisymmetry of f on $\mathbb{R}^{(n-1)\times(n-1)}$ together with the recursive definition (1.4) gives

$$f(A_1,\cdots,A_j,\cdots,A_k,\cdots,A_n)=0.$$

If $j = 1$, then by antisymmetry of f with respect to $A_2, \cdots, A_n$, we may assume without loss of generality that $k = 2$ and $A_1 = A_2$. For each $l = 3, \cdots, n$, we denote by $A_l^{(i)(j)} \in \mathbb{R}^{n-2}$ the vector obtained by deleting the i-th and j-th components of A_l. Applying (1.4) twice yields

$$f(A_1, A_2, \cdots, A_n) = \sum_{i=1}^{n} \sum_{j=1}^{n} (A_{i1} A_{j2} - A_{j1} A_{i2}) f(A_3^{(i)(j)}, \cdots, A_n^{(i)(j)}) = 0,$$

since $A_1 = A_2$. Hence, f is antisymmetric in all its arguments $A_1, A_2, \cdots, A_n$. This completes the proof. □

The following lemma establishes the uniqueness of the determinant function.

Lemma 1.2 *If f is a normalized, antisymmetric, and multilinear function on $\mathbb{R}^{n \times n}$, then we have*

$$f(A_1, \cdots, A_n) = \sum_{(i_1, \cdots, i_n) \in \mathcal{P}_n} s(i_1, \cdots, i_n) \prod_{j=1}^{n} A_{i_j j}, \tag{1.5}$$

where $\mathcal{P}_n$ denotes the set of all permutations of the indices $(1, 2, \cdots, n)$.

Proof By the multilinearity of f, we obtain

$$f(A_1, \cdots, A_n) = \sum_{(i_1, \cdots, i_n) \in \mathcal{P}_n} f(e_{i_1}, \cdots, e_{i_n}) \prod_{j=1}^{n} A_{i_j j}.$$

Since f is normalized and antisymmetric, it follows that

$$f(e_{i_1}, \cdots, e_{i_n}) = s(i_1, \cdots, i_n),$$

where $s(i_1, \cdots, i_n)$ denotes the sign of the permutation $(i_1, \cdots, i_n)$. This completes the proof. □

According to Lemma (1.1) and Lemma (1.2), the determinant function is uniquely determined by the conditions of normalization, antisymmetry, and multilinearity. From now on, we shall use det to denote the determinant function. The following properties are easy to verify.

Proposition 1.1 *Given $A = (A_1, \cdots, A_n) \in \mathbb{R}^{n \times n}$, we have*

$$\det(A) = \det(A^T). \tag{1.6}$$

For any $i, j = 1, \cdots, n$, we have

$$\sum_{k=1}^{n}(-1)^{i+k}A_{ki}\det(A^{(k,j)}) = \det(A)\delta_{ij} = \begin{cases}\det(A), & i = j,\\ 0, & i \neq j.\end{cases} \tag{1.7}$$

For any $\alpha \in \mathbb{R}$, *we have*

$$\det(A_1, \cdots, A_j, \cdots, A_k + \alpha A_j, \cdots, A_n) = \det(A_1, \cdots, A_n), \tag{1.8}$$

$$\det(A_1, \cdots, A_{k-1}, \alpha A_k, A_{k+1}, \cdots, A_n) = \alpha \det(A_1, \cdots, A_n). \tag{1.9}$$

For any $A, B \in \mathbb{R}^{n\times n}$, *we have*

$$\det(AB) = \det(A)\det(B). \tag{1.10}$$

In particular, if A is invertible, then $\det(A) \neq 0$. *On the other hand, if* $\det(A) \neq 0$, *then A is invertible with*

$$(A^{-1})_{ij} = \frac{(-1)^{i+j}\det(A^{(j,i)})}{\det(A)}. \tag{1.11}$$

1.3 Block Matrix

If $A \in \mathbb{R}^{m\times n}$, $B \in \mathbb{R}^{m\times q}$, $C \in \mathbb{R}^{p\times n}$, and $D \in \mathbb{R}^{p\times q}$, we define the *block matrix*

$$M = \begin{pmatrix} A & B \\ C & D \end{pmatrix} \in \mathbb{R}^{(m+p)\times(n+q)}$$

such that

$$M_{ij} = \begin{cases} A_{ij}, & i = 1, \cdots, m, \ j = 1, \cdots, n, \\ B_{i,j-n}, & i = 1, \cdots, m, \ j = n+1, \cdots, n+q, \\ C_{i-m,j}, & i = m+1, \cdots, m+p, \ j = 1, \cdots, n, \\ D_{i-m,j-n}, & i = m+1, \cdots, m+p, \ j = n+1, \cdots, n+q. \end{cases}$$

The operations of addition, subtraction, and scalar multiplication on block matrices can be carried out blockwise. The following lemma presents the formula for block matrix multiplication.

Lemma 1.3 *If* $A \in \mathbb{R}^{m\times n}$, $B \in \mathbb{R}^{m\times q}$, $C \in \mathbb{R}^{p\times n}$, $D \in \mathbb{R}^{p\times q}$, $\tilde{A} \in \mathbb{R}^{n\times r}$, $\tilde{B} \in \mathbb{R}^{n\times s}$, $\tilde{C} \in \mathbb{R}^{q\times r}$, *and* $\tilde{D} \in \mathbb{R}^{q\times s}$, *then*

$$\begin{pmatrix} A & B \\ C & D \end{pmatrix}\begin{pmatrix} \tilde{A} & \tilde{B} \\ \tilde{C} & \tilde{D} \end{pmatrix} = \begin{pmatrix} A\tilde{A} + B\tilde{C} & A\tilde{B} + B\tilde{D} \\ C\tilde{A} + D\tilde{C} & C\tilde{B} + C\tilde{D} \end{pmatrix}. \tag{1.12}$$

Proof Let

$$M = \begin{pmatrix} A & B \\ C & D \end{pmatrix} \in \mathbb{R}^{(m+p)\times(n+q)}, \quad \tilde{M} = \begin{pmatrix} \tilde{A} & \tilde{B} \\ \tilde{C} & \tilde{D} \end{pmatrix} \in \mathbb{R}^{(n+q)\times(r+s)}.$$

Then

$$(M\tilde{M})_{ij} = \sum_{k=1}^{n+q} M_{ik}\tilde{M}_{kj}, \quad i = 1, \cdots, m+p, \; j = 1, \cdots, r+s.$$

For $i = 1, \cdots, m$ and $j = 1, \cdots, r$, we have

$$(M\tilde{M})_{ij} = \sum_{k=1}^{n} A_{ik}\tilde{A}_{kj} + \sum_{k=n+1}^{n+q} B_{i,k-n}\tilde{C}_{k-n,j} = (A\tilde{A} + B\tilde{C})_{ij}.$$

We check each block explicitly. For $i = 1, \cdots, m$ and $j = r+1, \cdots, r+s$, we have

$$(M\tilde{M})_{ij} = \sum_{k=1}^{n} A_{ik}\tilde{B}_{k,j-r} + \sum_{k=n+1}^{n+q} B_{i,k-n}\tilde{D}_{k-n,j-r} = (A\tilde{B} + B\tilde{D})_{i,j-r}.$$

For $i = m+1, \cdots, m+p$ and $j = 1, \cdots, r$, we have

$$(M\tilde{M})_{ij} = \sum_{k=1}^{n} C_{i-m,k}\tilde{A}_{kj} + \sum_{k=n+1}^{n+q} D_{i-m,k-n}\tilde{C}_{k-n,j} = (C\tilde{A} + D\tilde{C})_{i-m,j}.$$

For $i = m+1, \cdots, m+p$ and $j = r+1, \cdots, r+s$, we have

$$(M\tilde{M})_{ij} = \sum_{k=1}^{n} C_{i-m,k}\tilde{B}_{k,j-r} + \sum_{k=n+1}^{n+q} D_{i-m,k-n}\tilde{D}_{k-n,j-r} = (C\tilde{B} + D\tilde{D})_{i-m,j-r}.$$

This completes the proof. □

The following result is useful for computing the determinant of block matrices.

Lemma 1.4 *Let*

$$M = \begin{pmatrix} A & B \\ C & D \end{pmatrix} \in \mathbb{R}^{(m+n)\times(m+n)},$$

where $A \in \mathbb{R}^{m\times m}$, $B = 0 \in \mathbb{R}^{m\times n}$, $C \in \mathbb{R}^{n\times m}$, *and* $D \in \mathbb{R}^{n\times n}$. *Then*

$$\det(M) = \det \begin{pmatrix} A & 0 \\ C & D \end{pmatrix} = \det(A)\det(D). \tag{1.13}$$

Proof We prove by induction on m. For $m = 1$, the result is immediate. Assume the formula holds for $m - 1$. For each $j = 1, \cdots, m$, we denote by $M^{(1,j)} \in \mathbb{R}^{(m+n-1)\times(m+n-1)}$ the submatrix obtained by deleting the first row and the j-th column of M:

$$M^{(1,j)} = \begin{pmatrix} A^{(1,j)} & 0 \\ C^{(0,j)} & D \end{pmatrix},$$

where $A^{(1,j)} \in \mathbb{R}^{(m-1)\times(m-1)}$ is obtained from A by removing its first row and j-th column, and $C^{(0,j)} \in \mathbb{R}^{n\times(m-1)}$ is obtained from C by removing its j-th column. By the induction hypothesis,

$$\det(M^{(1,j)}) = \det(A^{(1,j)})\det(D).$$

Expanding $\det(M)$ along its first row gives

$$\begin{aligned}\det(M) &= \sum_{j=1}^{m} (-1)^{j+1} A_{1j} \det(M^{(1,j)}) \\ &= \sum_{j=1}^{m} (-1)^{j+1} A_{1j} \det(A^{(1,j)})\det(D) \\ &= \det(A)\det(D),\end{aligned}$$

which completes the proof. □

Proposition 1.2 *Let*

$$M = \begin{pmatrix} A & B \\ C & D \end{pmatrix} \in \mathbb{R}^{2n\times 2n},$$

where $A, B, C, D \in \mathbb{R}^{n\times n}$ and $AC = CA$. Then

$$\det(M) = \det \begin{pmatrix} A & B \\ C & D \end{pmatrix} = \det(AD - CB). \tag{1.14}$$

Proof We first consider the case where A is invertible; that is, $\det(A) \neq 0$. Observe that

$$\begin{pmatrix} I & 0 \\ -CA^{-1} & I \end{pmatrix} \begin{pmatrix} A & B \\ C & D \end{pmatrix} = \begin{pmatrix} A & B \\ 0 & D - CA^{-1}B \end{pmatrix}.$$

Taking determinants on both sides and applying Proposition 1.1 and Lemma 1.4 give

$$\det\begin{pmatrix} A & B \\ C & D \end{pmatrix} = \det(A)\det(D - CA^{-1}B) = \det(AD - ACA^{-1}B).$$

Since $AC = CA$, it follows that $\det(M) = \det(AD - CB)$.

Next, assume that $\det(A) = 0$. Consider the polynomial

$$p(\lambda) = \det(A + \lambda I) = \lambda^n + \cdots, \quad \lambda \in \mathbb{R}.$$

By the fundamental theorem of algebra, $p(\lambda)$ has at most n real roots. Hence, there exists $\varepsilon_0 > 0$ such that $p(\varepsilon) \neq 0$ for all $\varepsilon \in (0, \varepsilon_0)$. For each such ε, set $A_\varepsilon = A + \varepsilon I$, which is invertible. By the previous argument, we have

$$\det\begin{pmatrix} A_\varepsilon & B \\ C & D \end{pmatrix} = \det(A_\varepsilon D - CB).$$

Letting $\varepsilon \to 0^+$ on both sides yields (1.14). This completes the proof. □

1.4 Rank

We say that the *rank* of a matrix $A \in \mathbb{R}^{m\times n}$ is r if there exist invertible matrices $P \in \mathbb{R}^{m\times m}$ and $Q \in \mathbb{R}^{n\times n}$ such that

$$PAQ = (e_1, \cdots, e_r, 0, \cdots, 0) \in \mathbb{R}^{m\times n},$$

where $e_1, \cdots, e_m$ are the standard basis vectors in $\mathbb{R}^m$. The following lemma ensures that the rank is well defined.

Lemma 1.5 *If*

$$A = P(e_1, \cdots, e_r, 0, \cdots, 0) = (e_1, \cdots, e_s, 0, \cdots, 0)Q \in \mathbb{R}^{m\times n},$$

where $P \in \mathbb{R}^{m\times m}$ and $Q \in \mathbb{R}^{n\times n}$ are invertible, then $r = s$.

Proof For any $i = 1, \cdots, m$, we have

$$A_{ij} = \begin{cases} P_{ij}, & j = 1, \cdots, r, \\ 0, & j = r+1, \cdots, n. \end{cases}$$

Similarly, for any $j = 1, \cdots, n$, we have

$$A_{ij} = \begin{cases} Q_{ij}, & i = 1, \cdots, s, \\ 0, & i = s+1, \cdots, m. \end{cases}$$

Combining these two formulas gives

$$\begin{aligned} P_{ij} =& 0, \ \ i = s+1, \cdots, m, \ \ j = 1, \cdots, r, \\ Q_{ij} =& 0, \ \ i = 1, \cdots, s, \ \ j = r+1, \cdots, n. \end{aligned}$$

Suppose $r < s$. We can write Q as a block matrix:

$$Q = \begin{pmatrix} B & 0 \\ C & D \end{pmatrix},$$

where $B = (B_1, \cdots, B_r, 0, \cdots, 0) \in \mathbb{R}^{s \times s}$, $C \in \mathbb{R}^{(n-s)\times s}$, and $D \in \mathbb{R}^{(n-s)\times(n-s)}$. By Lemma 1.4, we have $\det(Q) = \det(B)\det(D) = 0$, contradicting the invertibility of Q.

Suppose $r > s$. We can write P^T as a block matrix:

$$P^T = \begin{pmatrix} E & 0 \\ F & G \end{pmatrix},$$

where $E = (E_1, \cdots, E_s, 0, \cdots, 0) \in \mathbb{R}^{r \times r}$, $F \in \mathbb{R}^{(m-r)\times r}$, and $G \in \mathbb{R}^{(m-r)\times(m-r)}$. By Proposition 1.1 and Lemma 1.4, we have $\det(P) = \det(P^T) = \det(E)\det(G) = 0$, which contradicts the invertibility of P.

Hence $r = s$. This completes the proof. □

The following results are immediate consequences of the definition.

Proposition 1.3 *Let $A \in \mathbb{R}^{m\times n}$. Then*

$$\operatorname{rank}(A) = \operatorname{rank}(A^T) \le \min\{m, n\}.$$

Moreover, for any $A, B \in \mathbb{R}^{m\times n}$, we have $\operatorname{rank}(A) = \operatorname{rank}(B)$ if and only if there exist invertible matrices $P \in \mathbb{R}^{m\times m}$ and $Q \in \mathbb{R}^{n\times n}$ such that $A = PBQ$. In particular, a square matrix $A \in \mathbb{R}^{n\times n}$ is invertible if it has full rank, that is, $\operatorname{rank}(A) = n$.

We are now ready to show that matrix multiplication cannot increase the rank.

Proposition 1.4 *For any $A \in \mathbb{R}^{n\times m}$ and $B \in \mathbb{R}^{m\times p}$, we have*

$$\operatorname{rank}(AB) \le \min\{\operatorname{rank}(A), \operatorname{rank}(B)\}. \tag{1.15}$$

Proof For any matrix $M = P(e_1, \cdots, e_r, 0, \cdots, 0)Q$, where P and Q are invertible, we have

$$(M, 0) = P(e_1, \cdots, e_r, 0, \cdots, 0)\begin{pmatrix} Q & 0 \\ 0 & I \end{pmatrix},$$

and hence

$$\text{rank}(M, 0) = \text{rank}(M).$$

For any $R = (R_1, \cdots, R_m) \in \mathbb{R}^{n\times m}$ and any $s \leq m$, we have

$$\begin{aligned} &\text{rank}(R(e_1, \cdots, e_s, 0, \cdots, 0)) = \text{rank}(R_1, \cdots, R_s, 0, \cdots, 0) \\ &= \text{rank}(R_1, \cdots, R_s) \leq s. \end{aligned}$$

Now assume that $B = P(e_1, \cdots, e_s, 0, \cdots, 0)Q$, where P and Q are invertible. Then

$$\text{rank}(AB) = \text{rank}(AP(e_1, \cdots, e_s, 0, \cdots, 0)) \leq s = \text{rank}(B).$$

Furthermore

$$\text{rank}(AB) = \text{rank}((AB)^T) = \text{rank}(B^T A^T) \leq \text{rank}(A^T) = \text{rank}(A).$$

This completes the proof. □

The vectors $A_1, \cdots, A_k \in \mathbb{R}^n$ are said to be *linearly dependent* if there exists a nonzero vector $v \in \mathbb{R}^k$ such that

$$(A_1, \cdots, A_k)v = A_1v_1 + \cdots + A_kv_k = 0.$$

They are said to be *linearly independent* if the vector equation

$$(A_1, \cdots, A_k)v = 0$$

has only the trivial solution $v = 0 \in \mathbb{R}^k$.

Lemma 1.6 *The vectors $A_1, \cdots, A_k \in \mathbb{R}^n$ are linearly independent if and only if* $\text{rank}(A_1, \cdots, A_k) = k$.

Proof Suppose $(A_1, \cdots, A_k)v = 0$ for some nonzero vector $v \in \mathbb{R}^k$. Then there exists an index $i \in \{1, \cdots, k\}$ such that $v_i \neq 0$. Define

$$Q = (e_1, \cdots, e_{i-1}, e_{i+1}, \cdots, e_k, v) \in \mathbb{R}^{k\times k},$$

where $e_1, \cdots, e_k$ denote the standard basis vectors of $\mathbb{R}^k$. Since $\det(Q) = v_i \neq 0$, the matrix Q is invertible. Note that

$$(A_1, \cdots, A_k)Q = (A_1, \cdots, A_{i-1}, A_{i+1}, \cdots, A_k, 0),$$

which implies

$$\text{rank}(A_1, \cdots, A_k) = \text{rank}(A_1, \cdots, A_{i-1}, A_{i+1}, \cdots, A_k) \le k - 1 < k.$$

Conversely, if $\text{rank}(A_1, \cdots, A_k) = r < k$, then there exist invertible matrices $P = (P_1, \cdots, P_n) \in \mathbb{R}^{n\times n}$ and $Q = (Q_1, \cdots, Q_k) \in \mathbb{R}^{k\times k}$ such that

$$(A_1, \cdots, A_k)Q = P(e_1, \cdots, e_r, 0, \cdots, 0) = (P_1, \cdots, P_r, 0, \cdots, 0).$$

Let $v = Q_{r+1} \in \mathbb{R}^k$. Then $(A_1, \cdots, A_k)v = 0$, where $v \neq 0$ since $\det(Q) \neq 0$. Thus, the vectors $A_1, \cdots, A_k \in \mathbb{R}^n$ are linearly dependent. This completes the proof. □

Let $A = (A_1, \cdots, A_m) \in \mathbb{R}^{n\times m}$. Then $\text{rank}(A) = r$ if and only if there exist linearly independent vectors $P_1, \cdots, P_r \in \mathbb{R}^n$ such that each column $A_1, \cdots, A_m$ of A can be expressed as a linear combination of $P_1, \cdots, P_r$.

1.5 Fréchet Derivative

A function $f \in C(\mathbb{R}^n, \mathbb{R})$ is said to be *Fréchet differentiable* at $x \in \mathbb{R}^n$ if there exists a bounded linear operator $L \in C(\mathbb{R}^n, \mathbb{R})$ such that

$$\lim_{h\to 0} \frac{|f(x+h) - f(x) - Lh|}{\|h\|} = 0.$$

Any bounded linear operator $L \in C(\mathbb{R}^n, \mathbb{R})$ can be represented by a column vector $v \in \mathbb{R}^n$ satisfying $Lh = v^T h$ for all $h \in \mathbb{R}^n$. In this case, we denote $f'(x) = v \in \mathbb{R}^n$ and call $f'(x)$ the *gradient* of f at $x \in \mathbb{R}^n$. It follows that

$$f'(x) = \begin{pmatrix} \frac{\partial f}{\partial x_1}(x) \\ \cdots \\ \frac{\partial f}{\partial x_n}(x) \end{pmatrix} \in \mathbb{R}^n. \tag{1.16}$$

A vector-valued function $F \in C^1(\mathbb{R}^n, \mathbb{R}^m)$ is said to be *Fréchet differentiable* at $x \in \mathbb{R}^n$ if there exists a bounded linear operator $A \in C(\mathbb{R}^n, \mathbb{R}^m)$ such that

$$\lim_{h\to 0} \frac{|F(x+h) - F(x) - Ah|}{\|h\|} = 0.$$

We may regard the bounded linear operator $A \in C(\mathbb{R}^n, \mathbb{R}^m)$ as a matrix in $\mathbb{R}^{m\times n}$ and write $F'(x) = A^T \in \mathbb{R}^{n\times m}$. From the definition, we obtain

$$F'(x) = \begin{pmatrix} \frac{\partial F_1}{\partial x_1}(x) & \cdots & \frac{\partial F_m}{\partial x_1}(x) \\ \vdots & & \vdots \\ \frac{\partial F_1}{\partial x_n}(x) & \cdots & \frac{\partial F_m}{\partial x_n}(x) \end{pmatrix} \in \mathbb{R}^{n\times m}. \tag{1.17}$$

If $m = n$, we call $A = [F'(x)]^T$ the *Jacobian* matrix of F at $x \in \mathbb{R}^n$.

If $f \in C^2(\mathbb{R}^n, \mathbb{R})$, we define the *Hessian* matrix $f''(x) \in \mathbb{R}^{n\times n}$ as the Fréchet derivative of $f' \in C^1(\mathbb{R}^n, \mathbb{R}^n)$. In matrix form,

$$f''(x) = \begin{pmatrix} \frac{\partial^2 f}{\partial x_1 \partial x_1}(x) & \cdots & \frac{\partial^2 f}{\partial x_1 \partial x_n}(x) \\ \vdots & & \vdots \\ \frac{\partial^2 f}{\partial x_n \partial x_1}(x) & \cdots & \frac{\partial^2 f}{\partial x_n \partial x_n}(x) \end{pmatrix} \in \mathbb{R}^{n\times n}. \tag{1.18}$$

The following identities extend the product rule and chain rule for Fréchet derivatives.

Proposition 1.5 *Let $A \in \mathbb{R}^{n\times n}$, and define $f(x) = x^T Ax$ for $x \in \mathbb{R}^n$. Then*

$$f'(x) = (A + A^T)x \in \mathbb{R}^n, \quad f''(x) = A + A^T. \tag{1.19}$$

Let $F, G \in C^1(\mathbb{R}^n, \mathbb{R}^m)$, and define $f(x) = F(x)^T G(x)$ for $x \in \mathbb{R}^n$. Then

$$f'(x) = F'(x)G(x) + G'(x)F(x). \tag{1.20}$$

Given $A \in \mathbb{R}^{m\times p}$, $F \in C^1(\mathbb{R}^n, \mathbb{R}^m)$, and $G \in C^1(\mathbb{R}^n, \mathbb{R}^p)$, define $f(x) = F(x)^T AG(x)$ for $x \in \mathbb{R}^n$. We have

$$f'(x) = F'(x)AG(x) + G'(x)A^T F(x). \tag{1.21}$$

Given $F \in C^1(\mathbb{R}^m, \mathbb{R}^p)$ and $G \in C^1(\mathbb{R}^n, \mathbb{R}^m)$, define $H(x) = F(G(x))$ for $x \in \mathbb{R}^n$. We have

$$H'(x) = G'(x)F'(G(x)). \tag{1.22}$$

We are now ready to use Fréchet derivatives to state Taylor expansion in $\mathbb{R}^n$.

Proposition 1.6 *Assume that $f \in C^2(\mathbb{R}^n, \mathbb{R})$. For any $u, v \in \mathbb{R}^n$, there exists $w \in \mathbb{R}^n$ such that*

$$f(u) = f(v) + (u - v)^T f'(v) + \frac{1}{2}(u - v)^T f''(w)(u - v). \tag{1.23}$$

Proof Define

$$g(t) = f(v + (u - v)t), \quad t \in [0, 1].$$

By the chain rule in Proposition 1.5,

$$g'(t) = (u - v)^T f'(v + (u - v)t),$$

and

$$g''(t) = (u - v)^T f''(v + (u - v)t)(u - v).$$

Since $f \in C^2(\mathbb{R}^n, \mathbb{R})$, we have $g \in C^2(\mathbb{R}, \mathbb{R})$. By the one-dimensional Taylor's theorem, there exists $\xi \in (0, 1)$ such that

$$g(1) = g(0) + g'(0) + \frac{1}{2}g''(\xi).$$

Setting $w = v + (u - v)\xi \in \mathbb{R}^n$ gives (1.23), completing the proof. □

Problems

1.1 For any $A \in \mathbb{R}^{n\times m}$ and $B \in \mathbb{R}^{m\times p}$, prove that $(AB)^T = B^T A^T$.

1.2 If $A, B \in \mathbb{R}^{n\times n}$ are invertible, prove that AB is also invertible with $(AB)^{-1} = B^{-1}A^{-1}$.

1.3 Define $f(A) = A_{11}A_{22} - A_{21}A_{12}$ for $A \in \mathbb{R}^{2\times 2}$. Verify that f is a normalized, antisymmetric, and multilinear function on $\mathbb{R}^{2\times 2}$.

1.4 Show that a multilinear function f on $\mathbb{R}^{n\times n}$ is antisymmetric if and only if

$$f(A_1, \cdots, A_n) = 0$$

for any $A_1, \cdots, A_n \in \mathbb{R}^n$ with $A_j = A_k$ for some $j \neq k$.

1.5 Prove Proposition 1.1.

1.6 Given $\theta \in [0, \pi/2]$, evaluate $\det(A)$, where

$$A = \begin{pmatrix} 2\cos\theta & 1 & & & \\ 1 & 2\cos\theta & 1 & & \\ & \ddots & \ddots & \ddots & \\ & & 1 & 2\cos\theta & 1 \\ & & & 1 & 2\cos\theta \end{pmatrix} \in \mathbb{R}^{n\times n}.$$

Note that A is a tridiagonal matrix with $A_{ij} = \begin{cases} 2\cos\theta, & i = j, \\ 1, & |i - j| = 1, \\ 0, & |i - j| > 1. \end{cases}$

1.7 Evaluate $\det(B)$, where

$$B = \begin{pmatrix} 1 & 1 & \cdots & 1 \\ x_1 & x_2 & \cdots & x_n \\ \vdots & \vdots & & \vdots \\ x_1^{n-1} & x_2^{n-1} & \cdots & x_n^{n-1} \end{pmatrix} \in \mathbb{R}^{n\times n}.$$

Note that B is a Vandermonde matrix with $B_{ij} = x_j^{i-1}$.

1.8 For any $A \in \mathbb{R}^{m\times n}$ and $B \in \mathbb{R}^{n\times m}$, prove that $\det(I - AB) = \det(I - BA)$.

1.9 Prove Proposition 1.5.

Chapter 2
Linear Regression

Abstract In this chapter, we study the linear regression model as a fundamental example and a cornerstone of machine learning. Two equivalent approaches—maximum likelihood estimation and least squares approximation—are used to formulate the associated optimization problem. We derive the optimal solution and discuss its key properties. In addition, we introduce the variance inflation factor as a quantitative measure of linear correlation among features.

2.1 Notations

Throughout this chapter, we use d to denote the dimension (i.e., number of features) of the input data. The data size is n. The output data is a column vector

$$y = \begin{pmatrix} y_1 \\ y_2 \\ \vdots \\ y_n \end{pmatrix} \in \mathbb{R}^n. \tag{2.1}$$

To account for the constant term in the linear regression model, we introduce an additional column of ones and denote the augmented input data as

$$x = \begin{pmatrix} x_{10} & x_{11} & \cdots & x_{1d} \\ x_{20} & x_{21} & \cdots & x_{2d} \\ \vdots & \vdots & & \vdots \\ x_{n0} & x_{n1} & \cdots & x_{nd} \end{pmatrix} \in \mathbb{R}^{n\times(d+1)}, \tag{2.2}$$

where $x_{i0} = 1$ for $i = 1, \cdots, n$. For each $j = 0, \cdots, d$, we denote the j-th column of x by

X.-S. Wang, C. Wang, *Machine Learning in Data Processing*, Forum for Interdisciplinary Mathematics, https://doi.org/10.1007/978-3-032-20855-2_2

$$x_j = \begin{pmatrix} x_{1j} \\ x_{2j} \\ \vdots \\ x_{nj} \end{pmatrix} \in \mathbb{R}^n. \tag{2.3}$$

In particular,

$$x_0 = \begin{pmatrix} 1 \\ 1 \\ \vdots \\ 1 \end{pmatrix} \in \mathbb{R}^n \tag{2.4}$$

is the column vector added to the input data to represent the constant term in the linear regression model. The other column vectors $x_1, \cdots, x_d \in \mathbb{R}^n$ are also called the features of the input data, while the output vector $y \in \mathbb{R}^n$ is referred to as the observation.

2.2 Linear Model

The main problem in linear regression model is stated as follows, with illustration given in Fig. 2.1.

Objective
Given features $x_1, \cdots, x_d \in \mathbb{R}^n$ and observations $y \in \mathbb{R}^n$, we need to find a parameter vector

$$p = \begin{pmatrix} p_0 \\ p_1 \\ \vdots \\ p_d \end{pmatrix} \in \mathbb{R}^{d+1} \tag{2.5}$$

such that the prediction vector

$$Y := xp = x_0 p_0 + x_1 p_1 + \cdots + x_d p_d \tag{2.6}$$

is a good approximation of the observation vector y, where $x = (x_0, x_1, \cdots, x_d) \in \mathbb{R}^{n \times (d+1)}$ is the input data, see (2.2), and $x_0 \in \mathbb{R}^n$ is defined as in (2.4).

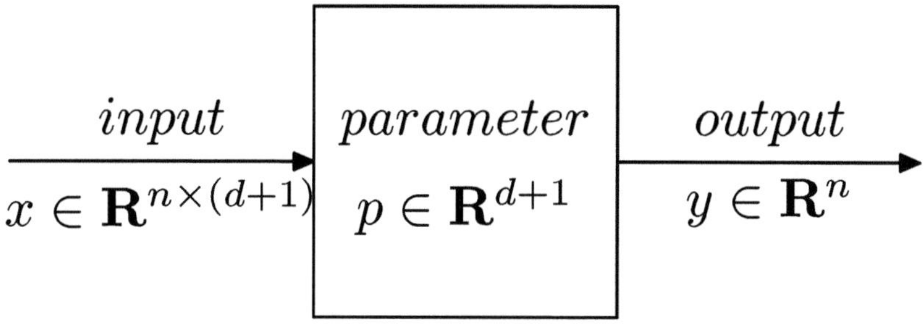

Fig. 2.1 Linear regression model of size n and dimension d

How to determine whether an approximation is good? There are two interpretations, which are essentially equivalent.

2.3 Maximum Likelihood Estimation

We assume that the error vector

$$\varepsilon := y - Y \in \mathbb{R}^n \tag{2.7}$$

satisfies a multivariate normal distribution with mean $0 \in \mathbb{R}^n$ and covariance $\Sigma = \sigma^2 I \in \mathbb{R}^{n\times n}$, where I stands for the identity matrix of dimension n. In other words, if

$$\varepsilon_i := y_i - Y_i = y_i - (xp)_i = y_i - \sum_{j=0}^{d} x_{ij} p_j \tag{2.8}$$

denotes the i-th component of the error vector (with $i = 1, \cdots, n$), then $\varepsilon_1, \cdots, \varepsilon_n$ are independent and identically distributed (i.i.d.) random variables satisfying a normal distribution with mean 0 and variance σ^2.

For a given observation vector $y \in \mathbb{R}^n$, the likelihood function is

$$L(p, \sigma^2 | y) = \frac{1}{(2\pi\sigma^2)^{n/2}} \exp\left\{ -\frac{1}{2\sigma^2} \sum_{i=1}^{n} \left(y_i - \sum_{j=0}^{d} x_{ij} p_j \right)^2 \right\}. \tag{2.9}$$

We then define a good approximation as the one that maximizes the above likelihood function. The problem of finding the maximum likelihood estimation of the parameter vector p is equivalent to the problem of minimizing the sum of squared errors (SSE) defined as below:

$$SSE := \sum_{i=1}^{n} \left(y_i - \sum_{j=0}^{d} x_{ij} p_j \right)^2 = \|y - Y\|_2^2, \tag{2.10}$$

where $\|\cdot\|_2$ denotes the l^2-norm in $\mathbb{R}^n$.

2.4 Least Squares Approximation

Another interpretation of the good approximation is the one that minimizes the l^2-norm of the error vector $y - Y$.

Optimization Problem
Let $x \in \mathbb{R}^{n\times(d+1)}$ and $y \in \mathbb{R}^n$ be given. Find $p \in \mathbb{R}^{d+1}$ such that the objective function

$$J(p) := \frac{1}{2}\|y - xp\|_2^2 = \frac{1}{2}\sum_{i=1}^{n} \left(y_i - \sum_{j=0}^{d} x_{ij} p_j \right)^2 \tag{2.11}$$

is minimized.

The additional factor $1/2$ in the objective function does not affect the solution of the optimization problem and is included only for the sake of convenience. The solution of the above optimization problem is also called the least squares approximation. Since the objective function $J(p)$ in (2.11) and the sum of squared errors (SSE) in (2.10) differ by a constant factor, the least squares approximation is equivalent to the maximum likelihood approximation.

Theorem 2.1 *If* $\det(x^T x) \neq 0$*, then the objective function* $J(p)$ *in* (2.11) *has a unique global minimum at*

$$p = (x^T x)^{-1}(x^T y); \tag{2.12}$$

namely, $J(q) > J(p)$ *for any* $q \in \mathbb{R}^{d+1}$ *and* $q \neq p$.

Proof Let p be given as in (2.12). For any $q \in \mathbb{R}^{d+1}$ with $q \neq p$, we consider the function

$$f(t) := J(tq + (1-t)p),$$

for $t \in \mathbb{R}$. Obviously, $f(0) = J(p)$ and $f(1) = J(q)$. By Taylor expansion, we obtain

$$f(1) = f(0) + f'(0) + \frac{1}{2} f''(s),$$

for some $s \in (0, 1)$. Note that

$$f'(0) = (q - p)^T J'(p) = (q - p)^T (x^T xp - x^T y) = 0,$$

and

$$f''(s) = (q-p)^T J''(sq+(1-s)p)(q-p) = (q-p)^T x^T x(q-p) = \|x(q-p)\|_2^2.$$

We obtain

$$J(q) - J(p) = f(1) - f(0) = \frac{1}{2} f''(s) = \frac{1}{2}\|x(q - p)\|_2^2 \geq 0.$$

Since $q - p \neq 0$ and $\det(x^T x) \neq 0$, we have $J(q) > J(p)$. Otherwise, $x(q - p) = 0$ and $x^T x(q-p) = 0$; namely, 0 is an eigenvalue of $x^T x$ with a corresponding eigenvector $q - p$, which is a contradiction to the assumption $\det(x^T x) \neq 0$. This completes the proof. □

2.5 Sum of Squared Errors

In this section, we assume $\det(x^T x) \neq 0$, and let p be the optimal solution as given in (2.12). The following proposition shows that the prediction vector $Y = xp$ is the projection of the observation vector y on the linear subspace spanned by the vectors $x_0, \cdots, x_d$; see Fig. 2.2.

Proposition 2.1 *Given* $x \in \mathbb{R}^{n\times(d+1)}$ *and* $y \in \mathbb{R}^n$ *with* $\det(x^T x) \neq 0$, *let* $Y = xp$ *with* $p = (x^T x)^{-1}(x^T y)$. *We have the following extension of Pythagorean theorem*

$$\|y - Y\|_2^2 = \|y\|_2^2 - \|Y\|_2^2. \tag{2.13}$$

Proof Note that

$$Y^T (y - Y) = p^T x^T y - p^T x^T xp = p^T (x^T y - x^T xp) = 0.$$

We have $Y^T Y = Y^T y = y^T Y$, and hence,

$$\|y - Y\|_2^2 = y^T y - Y^T y - y^T Y + Y^T Y = y^T y - Y^T Y = \|y\|_2^2 - \|Y\|_2^2.$$

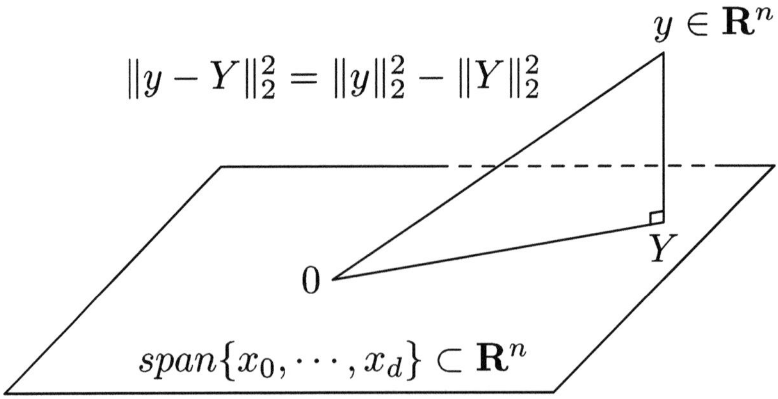

Fig. 2.2 Pythagorean theorem

This completes the proof. □

Denote

$$B := (x, y)^T (x, y) = \begin{pmatrix} x^T x & x^T y \\ y^T x & y^T y \end{pmatrix} \in \mathbb{R}^{(d+2)\times(d+2)}. \tag{2.14}$$

The following proposition shows that the last diagonal term (i.e., the $(d+2, d+2)$ entry) of B^{-1} is the reciprocal of the SSE.

Proposition 2.2 *Given $x \in \mathbb{R}^{n\times(d+1)}$ and $y \in \mathbb{R}^n$ with $\det(x^T x) \neq 0$, let $Y = xp$ with $p = (x^T x)^{-1}(x^T y)$. We have*

$$\begin{pmatrix} x^T x & x^T y \\ y^T x & y^T y \end{pmatrix}^{-1} = \begin{pmatrix} (x^T x)^{-1} + p\lambda p^T & -p\lambda \\ -\lambda p^T & \lambda \end{pmatrix}, \tag{2.15}$$

where

$$\lambda = \frac{1}{SSE} = \frac{1}{\|y - Y\|_2^2}.$$

Proof It is easily seen from $x^T x p = x^T y$ that

$$\begin{pmatrix} x^T x & x^T y \\ y^T x & y^T y \end{pmatrix} = \begin{pmatrix} x^T x & 0 \\ 0 & 1 \end{pmatrix} \begin{pmatrix} I & p \\ y^T x & y^T y \end{pmatrix}.$$

Assume

$$\begin{pmatrix} I & p \\ y^T x & y^T y \end{pmatrix} \begin{pmatrix} A & \alpha \\ \beta^T & \lambda \end{pmatrix} = \begin{pmatrix} I & 0 \\ 0 & 1 \end{pmatrix}.$$

We then have

$$A + p\beta^T = I, \;\; y^T x A + y^T y \beta^T = 0,$$
$$\alpha + p\lambda = 0, \;\; y^T x \alpha + y^T y \lambda = 1.$$

Solving these four equations yields

$$\alpha = -p\lambda, \;\; \lambda = \frac{1}{y^T y - y^T x p} = \frac{1}{y^T(y - Y)} = \frac{1}{SSE},$$

and

$$\beta^T = -\lambda y^T x, \;\; A = I + p\lambda y^T x.$$

Finally,

$$\begin{aligned}\begin{pmatrix} x^T x & x^T y \\ y^T x & y^T y \end{pmatrix}^{-1} &= \begin{pmatrix} I + p\lambda y^T x & -p\lambda \\ -\lambda y^T x & \lambda \end{pmatrix} \begin{pmatrix} (x^T x)^{-1} & 0 \\ 0 & 1 \end{pmatrix} \\ &= \begin{pmatrix} (x^T x)^{-1} + p\lambda y^T x (x^T x)^{-1} & -p\lambda \\ -\lambda y^T x (x^T x)^{-1} & \lambda \end{pmatrix} \\ &= \begin{pmatrix} (x^T x)^{-1} + p\lambda p^T & -p\lambda \\ -\lambda p^T & \lambda \end{pmatrix}.\end{aligned}$$

This completes the proof. □

2.6 Variance Inflation Factor

The sample variance of the output data is defined as

$$\sigma_y^2 := \frac{1}{n} \sum_{i=1}^{n} (y_i - \bar{y})^2, \tag{2.16}$$

where

$$\bar{y} := \frac{1}{n} \sum_{i=1}^{n} y_i \tag{2.17}$$

is the mean/average of the output data. Assume the input data $x \in \mathbb{R}^{n \times (d+1)}$ satisfies the condition $\det(x^T x) \neq 0$. Assume further that the sample variance of the output data is positive. We then define the coefficient of determination (R squared) of the linear regression model as

$$R^2 := 1 - \frac{SSE}{n\sigma_y^2} = 1 - \frac{\sum_{i=1}^n (y_i - Y_i)^2}{\sum_{i=1}^n (y_i - \bar{y})^2}, \tag{2.18}$$

where $SSE = \|y - Y\|_2^2$ and $Y = xp$ with $p = (x^T x)^{-1}(x^T y)$.

Theorem 2.2 *The coefficient of determination is bounded by* 1 *and nonnegative:* $0 \leq R^2 \leq 1$. *Moreover,* $R^2 = 1$ *if and only if the SSE is zero.*

Proof We only need to show that $R^2 \geq 0$. The other conclusions are obvious. Denote $q = (\bar{y}, 0, \cdots, 0)^T \in \mathbb{R}^{d+1}$. Recall the definition of the objective function in (2.11). It then follows from (2.18) that

$$R^2 = 1 - \frac{J(p)}{J(q)}.$$

By Theorem 2.1, we obtain $J(p) \leq J(q)$, and hence $R^2 \geq 0$. This completes the proof. □

Recall that $x = (x_0, x_1, \cdots, x_d) \in \mathbb{R}^{n\times(d+1)}$ with $x_0 = (1, 1, \cdots, 1)^T \in \mathbb{R}^n$. If $R^2 = 1$, then the observation vector y can be expressed as a linear combination of $x_0, \cdots, x_d$, and the matrix $B = (x, y)^T(x, y)$ becomes singular. For each $k = 1, \cdots, d$, we consider the linear regression model with $x^{(k)} := (x_0, \cdots, x_{k-1}, x_{k+1}, \cdots, x_d) \in \mathbb{R}^{n\times d}$ as the input data and x_k as the output data; that is, we find the parameter vector $p^{(k)} \in \mathbb{R}^d$ such that the prediction vector $X_k = x^{(k)}p^{(k)} \in \mathbb{R}^n$ has the smallest Euclidean distance from $x_k \in \mathbb{R}^n$. The corresponding sum of squared errors is denoted as

$$SSE_k := \|x_k - X_k\|_2^2 = \sum_{i=1}^n (x_{ik} - X_{ik})^2. \tag{2.19}$$

The sample variance of x_k is given by

$$\sigma_k^2 := \frac{1}{n}\sum_{i=1}^n (x_{ik} - \bar{x}_k)^2, \tag{2.20}$$

where

$$\bar{x}_k := \frac{1}{n}\sum_{i=1}^n x_{ik} \tag{2.21}$$

is the mean/average of x_k. Similar to the original linear regression model, we define the coefficient of determination (R squared) as

$$R_k^2 := 1 - \frac{SSE_k}{n\sigma_k^2} = 1 - \frac{\sum_{i=1}^n (x_{ik} - X_{ik})^2}{\sum_{i=1}^n (x_{ik} - \bar{x}_k)^2}. \tag{2.22}$$

Finally, the variance inflation factor (VIF) is defined as

$$VIF_k = \frac{1}{1 - R_k^2} = \frac{n\sigma_k^2}{SSE_k} = \frac{\sum_{i=1}^n (x_{ik} - \bar{x}_k)^2}{\sum_{i=1}^n (x_{ik} - X_{ik})^2}. \tag{2.23}$$

The following theorem indicates that a large VIF implies a strong linear correlation between the features.

Theorem 2.3 *Given the input data $x = (x_0, x_1, \cdots, x_d) \in \mathbb{R}^{n\times(d+1)}$ with $x_0 = (1, 1, \cdots, 1)^T \in \mathbb{R}^n$, let $VIF_1, \cdots, VIF_d$ be the variance inflation factors defined as in* (2.23). *We obtain* $\det(x^T x) = 0$ *if and only if $VIF_k = \infty$ for some $k \in \{1, \cdots, d\}$.*

Proof Note that $\det(x^T x) = 0$ if and only if there exists $k \in \{1, \cdots, d\}$ such that x_k is linearly dependent of $x_0, \cdots, x_{k-1}, x_{k+1}, \cdots, x_d$, which is also equivalent to $SSE_k = 0$, where SSE_k is the sum of squared errors defined in (2.19). On account of (2.23), we have $SSE_k = 0$ if and only if $VIF_k = \infty$. This completes the proof. □

The following proposition provides an alternative formula of the VIF.

Proposition 2.3 *Given the input data $x = (x_0, x_1, \cdots, x_d) \in \mathbb{R}^{n\times(d+1)}$ with $x_0 = (1, 1, \cdots, 1)^T \in \mathbb{R}^n$, assume* $\det(x^T x) \neq 0$, *and let $\lambda_0, \cdots, \lambda_d$ be the diagonal terms of $(x^T x)^{-1}$. For each $k = 1, \cdots, d$, let VIF_k be the variance inflation factor defined as in* (2.23). *We have $VIF_k = n\sigma_k^2 \lambda_k$, where σ_k^2 is the sample variance of x_k as given in* (2.20).

Proof By permutation, we may assume without loss of generality that $k = d$. A simple application of Proposition 2.2 gives $\lambda_k = 1/SSE_k$, where SSE_k is the sum of squared errors defined in (2.19). In view of (2.23), we obtain $VIF_k = n\sigma_k^2/SSE_k = n\sigma_k^2\lambda_k$. This completes the proof. □

2.7 Python Code

In this section, we provide a Python code to find the least squares solution, coefficient of determination, and variance inflation factors of the linear regression model.

Linear Regression Code

```
import numpy as np

def linear_regression(x,y):
    n=x.shape[0]
    p=np.dot(np.linalg.inv(np.dot(x.T,x)),np.dot(x.T,y))
    SSE=np.linalg.norm(y-np.dot(x,p))**2
    R2=1-SSE/(n*np.var(y))
    Lambda=np.diagonal(np.linalg.inv(np.dot(x.T,x)))
    VIF=np.multiply(Lambda,n*np.var(x,axis=0))[1:]
    result={
        'p':p,
        'R2':R2,
        'VIF':VIF
        }
    return result

n=1000
d=4
x0=np.ones((n,1))
x=np.concatenate((x0,np.random.random((n,d))),axis=1)
p_exact=np.ones((d+1,1))
y=np.dot(x,p_exact)+np.random.normal(0,0.1,(n,1))

result=linear_regression(x,y)
p=result['p']
R2=result['R2']
VIF=result['VIF']
print(R2)
print(VIF)
print(np.concatenate((p,p_exact),axis=1))
```

In the code, we choose the sample size $n = 1000$ and the dimension $d = 4$. The input data $x_1, \cdots, x_d \in \mathbb{R}^n$ is randomly generated from the uniform distribution. The exact parameter vector is set to be $(1, \cdots, 1)^T \in \mathbb{R}^{d+1}$. The output data is perturbed by normal random variables. A sample result is given below:

Sample Result

```
0.9702959227690222
[1.00333685 1.005423   1.00388619 1.00681191]
[[0.99675087 1.        ]
 [1.00069386 1.        ]
 [0.98648815 1.        ]
 [1.00547332 1.        ]
 [1.01346316 1.        ]]
```

It is noted that the coefficient of determination is very close to 1, the variance inflation factors are very close to 1, and the least squares solution is also very close to the exact solution.

To compute $p = (x^T x)^{-1}(x^T y)$, we adopt a code that, while being intuitively straightforward, regrettably suffers from low efficiency. Nonetheless, for data with a comparatively small dimension d, the utilization of this code is still tolerable. It must be stressed, however, that in actual practical applications, whenever feasible, one should avoid directly computing the inverse of $x^T x$. This is because such inverse computations generally involve substantial computational costs and complexity, inevitably resulting in poor efficiency. Hence, during real-world operations, it is essential to seek out more optimized computational pathways. In Chap. 9, we will introduce a collection of gradient methods to approximate the solution for the linear system $x^T xp = x^T y$, providing an alternative to circumvent the potential inefficiencies linked with direct computation of $(x^T x)^{-1}$ with large dimension d.

Problems

2.1 For any $x \in \mathbb{R}^{n\times(d+1)}$, show that $x^T x$ is a symmetric and positive semidefinite matrix in $\mathbb{R}^{(d+1)\times(d+1)}$. If $x^T x$ is invertible, then it is positive definite.

2.2 Let $x = (x_0, x_1, \cdots, x_d) \in \mathbb{R}^{n\times(d+1)}$, where $x_0, \cdots, x_d \in \mathbb{R}^n$ and $x_0 \neq 0$. Prove that $\det(x^T x) = 0$ if and only if there exists $k \in \{1, \cdots, d\}$ such that x_k can be represented as a linear combination of $x_0, \cdots, x_{k-1}, x_{k+1}, \cdots, x_d$; namely,

$$x_k = c_0 x_0 + \cdots + c_{k-1}x_{k-1} + c_{k+1}x_{k+1} + \cdots + c_d x_d,$$

for some $c_0, \cdots, c_{k-1}, c_{k+1}, \cdots, c_d \in \mathbb{R}$.

2.3 Given $x \in \mathbb{R}^{n\times(d+1)}$ and $y \in \mathbb{R}^n$, define a function

$$J(p) = \frac{1}{2}\|y - xp\|_2^2 = \frac{1}{2}\sum_{i=1}^{n}\left(y_i - \sum_{j=0}^{d} x_{ij}p_j\right)^2$$

for

$$p = \begin{pmatrix} p_0 \\ p_1 \\ \vdots \\ p_d \end{pmatrix} \in \mathbb{R}^{d+1}.$$

(a) Verify that $J'(p) = x^T(xp - y) \in \mathbb{R}^{d+1}$; namely,

$$\frac{\partial}{\partial p_k} J(p) = [J'(p)]_k = [x^T(xp - y)]_k = \sum_{i=1}^{n} x_{ik} \left(\sum_{j=0}^{d} x_{ij} p_j - y_i \right),$$

for $k = 0, \cdots, d$.

(b) Verify that $J''(p) = x^T x \in \mathbb{R}^{(d+1)\times(d+1)}$; namely,

$$\frac{\partial^2}{\partial p_j \partial p_k} J(p) = [J''(p)]_{jk} = (x^T x)_{jk} = \sum_{i=1}^{n} x_{ij} x_{ik},$$

for $j, k = 0, \cdots, d$.

2.4 Given $J \in C^2(\mathbb{R}^{d+1}, \mathbb{R})$, for any $p \in \mathbb{R}^{d+1}$ and $q \in \mathbb{R}^{d+1}$, prove that there exists $\xi \in \mathbb{R}^{d+1}$ such that

$$J(q) = J(p) + (q - p)^T J'(p) + \frac{1}{2}(q - p)^T J''(\xi)(q - p).$$

2.5 Prove that the variance inflation factor is no less than one; namely, $VIF_k \geq 1$, where VIF_k is defined in (2.23).

2.6 Given the input data $x = (x_0, x_1, \cdots, x_d) \in \mathbb{R}^{n\times(d+1)}$ with $x_0 = (1, 1, \cdots, 1)^T \in \mathbb{R}^n$, let SSE_k with $k = 1, \cdots, d$ be the sum of squared errors defined as in (2.23). Assume $\det(x^T x) \neq 0$, and let $\lambda_0, \cdots, \lambda_d$ be the diagonal terms of $(x^T x)^{-1}$. Show that $\lambda_k = 1/SSE_k$ for $k = 1, \cdots, d$.

2.7 Use a linear regression model to solve the problem of fitting the points $(0, 0)$, $(1, 2)$, $(2, 1)$, $(3, 3)$ by a straight line.

2.8 Given a positive integer n, calculate the mean and variance of the sequence $1, 2, \cdots, n$.

2.9 Given an input matrix

$$x = \begin{pmatrix} 1 & 1 & 0 \\ 1 & 0 & 1 \\ 1 & 0 & 0 \end{pmatrix},$$

calculate the variance inflation factors as defined in (2.23). Note that the sample size is $n = 3$ and the dimension is $d = 2$.

2.10 Given an input matrix $x \in \mathbb{R}^{n\times(d+1)}$ such that $\det(x^T x) \neq 0$, fix an observation vector $y \in \mathbb{R}^n$, and find the parameter vector $p \in \mathbb{R}^{d+1}$ and the variance $\sigma^2 > 0$ such that the likelihood function $L(p, \sigma^2|y)$ defined in (2.9) is maximized.

Chapter 3
Regularization

Abstract A large variance inflation factor in a linear regression model indicates strong linear correlation among features and may lead to overfitting. In this chapter, we address this issue by regularizing the objective function with the l^2-norm of the parameter vector and introduce the ridge regression model. We also study the LASSO (least absolute shrinkage and selection operator) method, which employs l^1-regularization. Since the l^1-norm is not differentiable, we introduce the concept of the subdifferential and formulate the problem within the framework of convex optimization. An iterative algorithm is then derived to compute a numerical solution for linear regression with l^1-regularization.

3.1 Overfitting Problem

We continue with the linear regression code in Sect. 2.7. Now, we assume x_2 and x_1 are closely related in the sense that the difference $x_2 - x_1$ is sampled from a normal distribution with mean 0 and a small variance. For simplicity, we do not repeat the code for the linear regression function in Sect. 2.7.

Overfitting Example

```
import numpy as np

n=1000
d=4
x0=np.ones((n,1))
x=np.concatenate((x0,np.random.random((n,d))),axis=1)
x[:,2]=x[:,1]+np.random.normal(0,0.01,(n,))
p_exact=np.ones((d+1,1))
y=np.dot(x,p_exact)+np.random.normal(0,0.1,(n,1))
```

(continued)

X.-S. Wang, C. Wang, *Machine Learning in Data Processing*, Forum for Interdisciplinary Mathematics, https://doi.org/10.1007/978-3-032-20855-2_3

```
result=linear_regression(x,y)
p=result['p']
R2=result['R2']
VIF=result['VIF']
print(R2)
print(VIF)
print(np.concatenate((p,p_exact),axis=1))
```

A sample result is given below:

Overfitting Result

```
0.9796829459418794
[803.11874709 802.97720154    1.01312437    1.01804729]
[[1.00094331 1.        ]
 [1.19645218 1.        ]
 [0.7916922  1.        ]
 [1.00670459 1.        ]
 [1.0056871  1.        ]]
```

It is noted that the coefficient of determination is still close to 1. However, the variance inflation factors VIF_1 and VIF_2 are large. This indicates that the features may be linearly correlated and the model may encounter a problem of overfitting. The least squares approximation shows that p_1 and p_2 are not close to 1, while the sum $p_1 + p_2$ is very close to $1 + 1 = 2$.

In practice, when we deal with a high-dimension problem with many features, it is very likely that some features are linearly correlated. In this case, we may find many solutions with small bias (i.e., the SSE is small). In particular, if $\det(x^T x) = 0$, then the linear regression model has infinitely many solutions. Moreover, when the variance is high (i.e., the VIF is large), the condition number of $x^T x$ becomes large, and the solution of $x^T xp = x^T y$ has low accuracy. To balance the bias and variance, we need to introduce the l^2-regularization and consider the ridge regression problem.

3.2 Ridge Regression

In the case of large VIF (i.e., $x^T x$ is ill-conditioned), we need to modify the optimization problem in Sect. 2.4 with the following objective function:

$$J(p) = \frac{1}{2}\|y - xp\|_2^2 + \frac{\lambda}{2}\|p\|_2^2, \tag{3.1}$$

where $\lambda > 0$ is a regularization constant to be tuned. It is also called a *hyperparameter* or a tuning parameter.

Theorem 3.1 *For any $x \in \mathbb{R}^{n\times(d+1)}$, $y \in \mathbb{R}^n$, and $\lambda > 0$, the objective function $J(p)$ in* (3.1) *has a unique global minimum at*

$$p = (x^T x + \lambda I)^{-1}(x^T y); \tag{3.2}$$

namely, $J(q) > J(p)$ for any $q \in \mathbb{R}^{d+1}$ and $q \neq p$.

Proof Let p be given as in (3.2). We have

$$J'(p) = x^T(xp - y) + p\lambda = (x^T x + \lambda I)p - x^T y = 0.$$

For any $q \in \mathbb{R}^{d+1}$ with $q \neq p$, we obtain from Taylor expansion

$$J(q) - J(p) = \frac{1}{2}(q - p)^T(x^T x + \lambda I)(q - p) \geq \frac{\lambda}{2}\|(q - p)\|_2^2 > 0.$$

This completes the proof. □

We come back to the overfitting problem in Sect. 3.1. The Python code implementing the ridge regression is given below:

Ridge Regression

```
import numpy as np

n=1000
d=4
x0=np.ones((n,1))
x=np.concatenate((x0,np.random.random((n,d))),axis=1)
x[:,2]=x[:,1]+np.random.normal(0,0.01,(n,))
p_exact=np.ones((d+1,1))
y=np.dot(x,p_exact)+np.random.normal(0,0.1,(n,1))
```

(continued)

```
p=np.dot(np.linalg.inv(np.dot(x.T,x)),np.dot(x.T,y))
la=1
A=la*np.eye(d+1)+np.dot(x.T,x)
p_ridge=np.dot(np.linalg.inv(A),np.dot(x.T,y))
print(np.concatenate((p,p_exact,p_ridge),axis=1))
```

We choose the regularization parameter $\lambda = 1$. A sample result is given below:

Sample Result

```
[[0.98693898 1.          0.99435015]
 [1.15005244 1.          1.01000321]
 [0.86133075 1.          0.99605137]
 [1.00696983 1.          1.00073844]
 [0.99278865 1.          0.98752897]]
```

It is noted that the result from the ridge regression (linear regression with l^2-regularization) is closer to the exact parameter vector than that of linear regression. This example illustrates that l^2-regularization helps reduce the side effect of overfitting the data with correlated features.

3.3 Convex Optimization

Another commonly used regularization method is adding the l^1-norm of the parameter vector to the objective function. As we shall see later, this regularization helps reduce the number of nonzero parameters in the estimation and enables us to select the significant features of the input data. The method is also called LASSO, which stands for least absolute shrinkage and selection operator.

Since l^1-norm may not be differentiable at some points, we need to generalize the definitions of derivatives and the ideas of critical points. First, we recall the definition of a convex function in an interval. Let $I \subset \mathbb{R}$ be an open interval. A continuous function $f \in C(I, \mathbb{R})$ is *c*onvex if and only if

$$f(tu + (1-t)v) \leq tf(u) + (1-t)f(v), \tag{3.3}$$

for all $u \in I$, $v \in I$, and $t \in [0, 1]$. Note that a convex function is not necessarily differentiable. For example, $f(u) = |u|$ is convex in $\mathbb{R}$ but not differentiable at $u = 0$. The following lemma gives an alternative definition of a convex function; see Fig. 3.1.

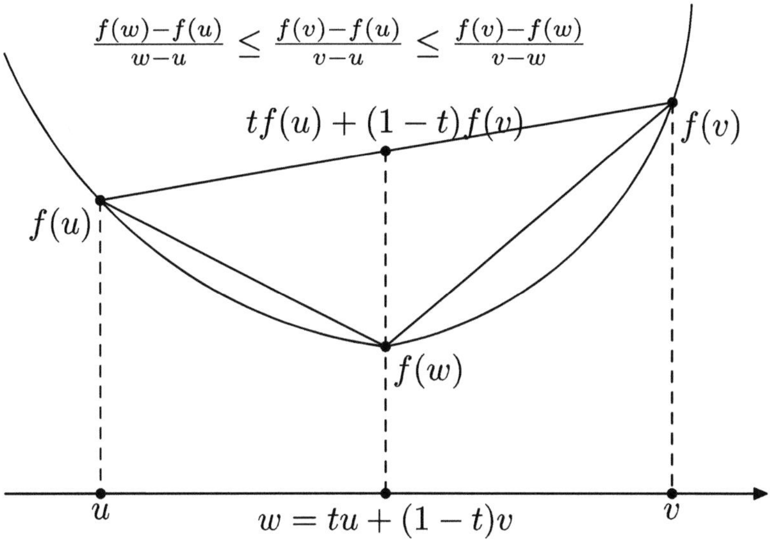

Fig. 3.1 Convex function

Lemma 3.1 *Let $I \subset \mathbb{R}$ be an open interval and $f \in C(I, \mathbb{R})$. Then f is convex in I if and only if for any $u \in I$, $v \in I$, and $w \in I$ such that $u < w < v$, we have*

$$\frac{f(w) - f(u)}{w - u} \le \frac{f(v) - f(u)}{v - u} \le \frac{f(v) - f(w)}{v - w}. \tag{3.4}$$

Proof First, we assume that (3.4) holds for all $u \in I$, $v \in I$, and $w \in I$ such that $u < w < v$. In particular, by choosing $w = tu + (1-t)v$ with $t \in (0, 1)$, we obtain from the first inequality in (3.4) that

$$\frac{f(w) - f(u)}{(1-t)(v-u)} \le \frac{f(v) - f(u)}{v - u}.$$

A further simplification yields (3.3). When $t = 0$ or $t = 1$, the inequality (3.3) becomes an obvious equality.

Next, we assume that (3.3) holds for all $u \in I$, $v \in I$, and $t \in [0, 1]$. Given any $w \in (u, v)$, we choose $t = (v-w)/(v-u) \in (0, 1)$ such that $w = tu+(1-t)v$. The inequalities in (3.4) can be easily obtained from (3.3). This completes the proof. □

Let $f \in C(I, \mathbb{R})$ be a convex function on an open interval $I \subset \mathbb{R}$. For any $u \in I$, one can prove that the one-sided derivatives

$$f' \pm (u) := \lim_{h \to 0^{\pm}} \frac{f(u+h) - f(u)}{h}$$

exist and $f' - (u) \le f' + (u)$. We then define the subdifferential of f at u as the following closed interval:

$$\partial f(u) := [f' - (u), f' + (u)]. \tag{3.5}$$

Any value in $\partial f(u)$ is called a subderivative of f at u. If f is differentiable at u, then the subdifferential shrinks to a single point $f'(u)$, and the subderivative coincides with the derivative. It is well known that if f is convex and differentiable on an open interval $I \subset \mathbb{R}$, then $u \in I$ is a local minimum of f if and only if $f'(u) = 0$. Moreover, any local minimum of f in I is global in I. By using the subdifferential, we can extend this result in the following proposition.

Proposition 3.1 *Let $f \in C(I, \mathbb{R})$ be a convex function on an open interval $I \subset \mathbb{R}$. We have the following results:*

(a) $u \in I$ is a local minimum point of f if and only if $0 \in \partial f(u)$.
(b) If $u \in I$ and $v \in I$ with $u < v$ are two local minimum points of f, then $f'(w) = 0$ and $f(w) = f(u) = f(v)$ for any $w \in (u, v)$.

Proof If $u \in I$ is a local minimum point of f, then $f(u + h) \ge f(u)$ for any sufficiently small $h > 0$, and hence $f' + (u) \ge 0$. Similarly, we have $f' - (u) \le 0$. Therefore, $0 \in \partial f(u)$. On the other hand, if $0 \in \partial f(u)$, namely, $f' - (u) \le 0 \le f' + (u)$, then we have

$$\lim_{h \to 0^+} g(h) \ge 0,$$

where $g(h) := [f(u + h) - f(u)]/h$ is the difference quotient. By Lemma 3.1, $g(h_1) \ge g(h_2)$ for any small $h_1 > h_2 > 0$. Hence, $g(h) \ge 0$ and $f(u+h) \ge f(u)$ for all small $h > 0$. Similarly, we can show that $f(u - h) \ge f(u)$ for all small $h > 0$. Therefore, u is a local minimum of f. This proves (a).

Now, we assume that $u \in I$ and $v \in I$ with $u < v$ are two local minimum points of f. By (a), we have

$$f' + (u) \ge 0 \ge f' - (v).$$

By letting $w \to u^+$ in the first inequality of (3.4) and $w \to v^-$ in the second inequality of (3.4), respectively, we obtain

$$f' + (u) \le \frac{f(v) - f(u)}{v - u} \le f' - (v).$$

Coupling the above two inequalities gives

$$f' + (u) = f' - (v) = f(v) - f(u) = 0.$$

Moreover, for any $w \in (u, v)$, we obtain from

$$0 = f' + (u) \leq f' - (w) \leq f' + (w) \leq f' - (v) = 0$$

that $f'(w) = 0$. A simple integration from u to w then gives $f(w) = f(u) = f(v)$ for all $w \in (u, v)$. This proves (b). □

Denote $f(u) := \lambda|u|$. A simple calculation gives

$$\partial f(u) = \begin{cases} \lambda, & u > 0, \\ -\lambda, & u < 0, \\ [-\lambda, \lambda], & u = 0. \end{cases} \tag{3.6}$$

Consider a simple optimization problem with the objective function

$$J(u) = \frac{1}{2}(u - z)^2 + \lambda|u|, \quad u \in \mathbb{R}, \tag{3.7}$$

where $z \in \mathbb{R}$ and $\lambda > 0$ are given. Denote

$$F_\lambda(z) := \begin{cases} z - \lambda, & z > \lambda, \\ z + \lambda, & z < -\lambda, \\ 0, & z \in [-\lambda, \lambda]. \end{cases} \tag{3.8}$$

Lemma 3.2 *The objective function $J(u)$ in* (3.7) *is minimized at $u^* = F_\lambda(z)$ as defined in* (3.8).

Proof By Proposition 3.1, $u^* \in \mathbb{R}$ is a local minimum of $J(u)$ if and only if $z - u^* \in \partial f(u^*)$. In view of (3.6), we obtain (i) $z - u^* = \lambda$ if $u^* > 0$, (ii) $z - u^* = -\lambda$ if $u^* < 0$, and (iii) $z - u^* \in [-\lambda, \lambda]$ if $u^* = 0$. This proves $u^* = F_\lambda(z)$. □

The following theorem shows that the local minimum of a sum of $f(u) = \lambda|u|$ with $\lambda > 0$ and a differentiable function g can be interpreted as a fixed point of a function involving F_λ defined in (3.8).

Theorem 3.2 *Let $g \in C^1(I, \mathbb{R})$ be differentiable in an open interval $I \subset \mathbb{R}$. Set $f(u) = \lambda|u|$ with $\lambda > 0$. Then $u^* \in I$ is a local minimum of $f(u) + g(u)$ if and only if $u^* = F_{\lambda\alpha}(u^* - \alpha g'(u^*))$ for any $\alpha > 0$. Here, F_λ is defined in* (3.8).

Proof Note that $u^* \in I$ is a local minimum of $f(u) + g(u)$ if and only if $0 \in \partial f(u^*) + g'(u^*)$. For any $\alpha > 0$, we also note that $u^* \in I$ is a local minimum of $\alpha f(u) + [u - u^* + \alpha g'(u^*)]^2/2$ if and only if $0 \in \alpha\partial f(u^*) + \alpha g'(u^*)$. Hence, $u^* \in I$ is a local minimum of $f(u) + g(u)$ if and only if it is a local minimum of $\alpha f(u) + [u - u^* + \alpha g'(u^*)]^2/2$ for any $\alpha > 0$, which satisfies the fixed point equation $u^* = F_{\lambda\alpha}(u^* - \alpha g'(u^*))$. This completes the proof. □

3.4 LASSO

We are now ready to investigate the optimization problem with l^1-regularization, which is also referred to as LASSO (least absolute shrinkage and selection operator).

LASSO
Let $x \in \mathbb{R}^{n\times(d+1)}$ and $y \in \mathbb{R}^n$ be given. Find $p \in \mathbb{R}^{d+1}$ such that the objective function

$$J(p) := \frac{1}{2}\|y - xp\|_2^2 + \lambda\|p\|_1 = \frac{1}{2}\sum_{i=1}^{n}\left(y_i - \sum_{j=0}^{d} x_{ij}p_j\right)^2 + \lambda\sum_{j=0}^{d}|p_j| \tag{3.9}$$

is minimized.

We shall extend Theorem 3.2 from one-dimensional real line to a higher dimensional space.

Theorem 3.3 *Let F_λ be defined as in* (3.8). *Given $x \in \mathbb{R}^{n\times(d+1)}$ and $y \in \mathbb{R}^n$, fix any $\alpha > 0$. Then $p \in \mathbb{R}^{d+1}$ is a local minimum of the objective function $J(p)$ in* (3.9) *if and only if*

$$p_j = F_{\lambda\alpha}(p_j - \alpha(x^T xp - x^T y)_j),$$

for $j = 0, \cdots, d$.

Proof Define

$$g(p) := \frac{1}{2}\|y - xp\|_2^2.$$

Note that $g'(p) = x^T(xp - y)$. For each $j = 0, \cdots, d$, we have

$$\frac{\partial}{\partial p_j}g(p) = (x^T xp - x^T y)_j = \sum_{i=1}^{n}\sum_{k=0}^{d} x_{ij}x_{ik}p_k - \sum_{i=1}^{n} x_{ij}y_i.$$

By regarding $g(p)$ as a one-variable function of p_j, we obtain from Theorem 3.2 our desired formula. This completes the proof. □

For simplicity, we extend the definition of F_λ in (3.8) to be a function in $C(\mathbb{R}^{d+1}, \mathbb{R}^{d+1})$ such that

$$[F_\lambda(u)]_j = F_\lambda(u_j) \tag{3.10}$$

for $u = (u_0, u_1, \cdots, u_d) \in \mathbb{R}^{d+1}$ and $j = 0, 1, \cdots, d$. Theorem 3.3 provides an iterative method of finding a local minimum of $J(p)$ in (3.9).

$$p^{(k+1)} = F_{\lambda\alpha}(p^{(k)} - \alpha x^T(xp^{(k)} - y)). \tag{3.11}$$

Here, $\lambda > 0$ is the regularization parameter, and $\alpha > 0$ is also called the *l*earning rate. Both hyperparameters are to be tuned in practical applications. In the special case $\lambda \to 0^+$, the above iteration reduces to the steepest descent method:

$$p^{(k+1)} = p^{(k)} - \alpha x^T(xp^{(k)} - y) = p^{(k)} - \alpha J'(p^{(k)}).$$

Comparing with the direct formula in (2.12), the steepest descent method avoids solving the inverse of the matrix $x^T x$ which is generally dense and sometimes ill-conditioned. The choice of the learning rate $\alpha > 0$ depends on the specific problem. In principle, a large α may lead to divergence of the iteration, while a small α may induce some computation costs.

3.5 Python Code

In this section, we provide a Python code to implement the iterative algorithm (3.11) in LASSO.

LASSO Code

```
import numpy as np

def F(u,la):
    return (u>la)*(u-la)+(u<-la)*(u+la)

def lasso(x,y,la,al,iter_max):
    d=x.shape[1]-1
    p=np.zeros((d+1,1))
    for iter in range(iter_max):
        u=p-al*np.dot(x.T,np.dot(x,p)-y)
        p=F(u,la*al)
    return p

n=1000
```

(continued)

```
d=4
x0=np.ones((n,1))
x=np.concatenate((x0,np.random.random((n,d))),axis=1)
p_exact=np.ones((d+1,1))
p_exact[3]=0
p_exact[4]=0
y=np.dot(x,p_exact)+np.random.normal(0,0.1,(n,1))

p=np.dot(np.linalg.inv(np.dot(x.T,x)),np.dot(x.T,y))
p_lasso=lasso(x,y,la=1,al=0.0001,iter_max=10000)
print(np.concatenate((p,p_exact,p_lasso),axis=1))
```

In the code, we choose the sample size $n = 1000$ and the dimension $d = 4$. The input data $x_1, \cdots, x_d \in \mathbb{R}^n$ is randomly generated from the uniform distribution. The exact parameter vector is set to be $p_{exact} = (1, 1, 1, 0, 0)^T \in \mathbb{R}^5$. The output data is perturbed by normal random variables. A sample result is given below:

Sample Result

```
[[ 1.01146268  1.          1.00860743]
 [ 1.00235203  1.          0.99627135]
 [ 0.99401163  1.          0.9872306 ]
 [-0.0103034   0.          0.        ]
 [-0.00607228  0.          0.        ]]
```

It is noted that the linear regression (without regularization) gives small but nonzero estimations to p_3 and p_4. However, by choosing a suitable regularization parameter, the LASSO method estimates $p_3 = 0$ and $p_4 = 0$. This example shows that the LASSO method is able to select the significant features of the input data.

3.6 Discussions

We use an illustrative example to compare l^2-regularization and l^1-regularization. Consider an objective function

$$J(p) = \frac{1}{2}(p_0 - 1)^2 + \frac{1}{2}(p_1 - 0.01)^2,$$

for $p = (p_0, p_1)^T \in \mathbb{R}^2$. Clearly, $J(p)$ is minimized at

$$p^* = \begin{pmatrix} 1 \\ 0.01 \end{pmatrix}.$$

Now, we fix $\lambda > 0$ and consider two objective functions

$$\hat{J}(p) = \frac{1}{2}(p_0 - 1)^2 + \frac{1}{2}(p_1 - 0.01)^2 + \frac{\lambda}{2}(p_0^2 + p_1^2),$$

and

$$\tilde{J}(p) = \frac{1}{2}(p_0 - 1)^2 + \frac{1}{2}(p_1 - 0.01)^2 + \lambda(|p_0| + |p_1|),$$

respectively, as the l^2-regularization and l^1-regularization of $J(p)$. It is easily seen that $\hat{J}(p)$ is minimized at

$$\hat{p}^* = \frac{1}{1+\lambda}\begin{pmatrix} 1 \\ 0.01 \end{pmatrix},$$

which approaches p^* as $\lambda \to 0$ and $(0, 0)^T$ as $\lambda \to \infty$. Finally, we obtain from Lemma 3.2 that $\tilde{J}(p)$ is minimized at

$$\tilde{p}^* = \begin{pmatrix} (1-\lambda)_+ \\ (0.01-\lambda)_+ \end{pmatrix}, \tag{3.12}$$

where the plus subscript denotes the maximum of the number and zero; namely,

$$x_+ := \max\{x, 0\} = \begin{cases} x, & x \geq 0, \\ 0, & x \leq 0. \end{cases} \tag{3.13}$$

Similar to l^2-regularization, we have $\tilde{p}^* \to p^*$ as $\lambda \to 0$ and $\tilde{p}^* \to (0, 0)^T$ as $\lambda \to \infty$. An important and useful property of l^1-regularization is that only the significant features are selected if the regularization parameter is appropriately chosen. In our illustrative example, we have $\tilde{p}_1 = 0$ for $\lambda \in [0.01, 1)$. It is necessary to point out that $\tilde{p}^* = (0, 0)^T$ if $\lambda \geq 1$. This suggests us that in practice, we should avoid setting λ to be too large; otherwise the parameter vector will become zero, and none of the features will be selected.

Problems

3.1 For any $x \in \mathbb{R}^{n\times(d+1)}$ and $y \in \mathbb{R}^n$, show that the equation $x^T x p = x^T y$ has infinitely many solutions if and only if $\det(x^T x) = 0$.

3.2 Given the input data

$$x = \begin{pmatrix} 1 & 0 & 1 \\ 1 & 1 & 2 \\ 1 & 2 & 3 \\ 1 & 3 & 4 \end{pmatrix}$$

and the output data $y = (0, 2, 1, 3)^T$, find the optimal parameter vector(s) $p \in \mathbb{R}^3$ such that the sum of squared errors $SSE = \|y - xp\|_2^2$ is minimized.

3.3 Given $x \in \mathbb{R}^{n\times(d+1)}$ and $\lambda > 0$, prove that the matrix $x^T x + \lambda I$ is positive definite and hence invertible, where I is the identity matrix of dimension $d + 1$.

3.4 Given $x \in \mathbb{R}^{n\times(d+1)}$, $y \in \mathbb{R}^n$, and $\lambda > 0$, define a function

$$J(p) = \frac{1}{2}\|y - xp\|_2^2 + \frac{\lambda}{2}\|p\|_2^2 = \frac{1}{2}\sum_{i=1}^{n}\left(y_i - \sum_{j=0}^{d} x_{ij}p_j\right)^2 + \frac{1}{2}\sum_{j=0}^{d} p_j^2$$

for

$$p = \begin{pmatrix} p_0 \\ \partial_1 \\ \vdots \\ \partial_d \end{pmatrix} \in \mathbb{R}^{d+1}.$$

(a) Verify that $J'(p) = x^T(xp - y) + p\lambda \in \mathbb{R}^{d+1}$; namely,

$$\frac{\partial}{\partial p_k} J(p) = [J'(p)]_k = \sum_{i=1}^{n} x_{ik}\left(\sum_{j=0}^{d} x_{ij}p_j - y_i\right) + \lambda p_k,$$

for $k = 0, \cdots, d$. (b) Verify that $J''(p) = x^T x + \lambda I \in \mathbb{R}^{(d+1)\times(d+1)}$; namely,

$$\frac{\partial^2}{\partial p_j \partial p_k} J(p) = [J''(p)]_{jk} = \sum_{i=1}^{n} x_{ij}x_{ik} + \lambda\delta_{jk},$$

for $j, k = 0, \cdots, d$, where

$$\delta_{jk} = \begin{cases} 1 & j = k, \\ 0 & j \neq k \end{cases}$$

is the Kronecker delta symbol.

3.5 In the Python code given in Section 3.2, tune the hyperparameter λ from 10^{-3} to 10^3, and find the pattern of changes in the bias and variance of the solution.

3.6 Let $f \in C^2(I, \mathbb{R})$, where $I \subset \mathbb{R}$ is an open interval. Prove that f is convex in I if and only if $f''(u) \geq 0$ for all $u \in I$.

3.7 Let $f \in C(I, \mathbb{R})$ be a convex function on an open interval $I \subset \mathbb{R}$. For any $u \in I$, prove that the one-sided derivatives

$$f' \pm (u) := \lim_{h \to 0^{\pm}} \frac{f(u+h) - f(u)}{h}$$

exist and $f' - (u) \leq f' + (u)$. Moreover, the derivative $f'(u)$ exists if and only if $f' - (u) = f' + (u)$.

3.8 Let $f \in C^1(I, \mathbb{R})$ be a convex function on an open interval $I \subset \mathbb{R}$. Then $u \in I$ is a local minimum of f if and only if $f'(u) = 0$. Moreover, any local minimum of f in I is global in I.

3.9 Let $f \in C(I, \mathbb{R})$ be a convex function on an open interval $I \subset \mathbb{R}$. For any $u \in I$ and $v \in I$ with $u < v$, prove that $f' + (u) \leq f' - (v)$.

3.10 Let $f \in C(I, \mathbb{R})$ be a convex function on an open interval $I \subset \mathbb{R}$. Given $u \in I$, prove that $c \in \partial f(u)$ if and only if $f(v) - f(u) \geq c(v - u)$ for any $v \in I$.

3.11 Let $f \in C(I, \mathbb{R})$ be a convex function on an open interval $I \subset \mathbb{R}$. Given $g \in C^1(I, \mathbb{R})$, prove that $u \in I$ is a local minimum of $f + g$ if and only if $0 \in \partial f(u) + g'(u)$ (i.e., $f' - (u) \leq -g'(u) \leq f' + (u)$).

3.12 Verify (3.12).

Chapter 4
Nonlinear Regression

Abstract Linear regression is applicable only when the output depends approximately linearly on the input. In many real-world datasets, however, this relationship is not necessarily linear, and a nonlinear regression model may provide a more appropriate description. In this chapter, we use logic gates as illustrative examples to demonstrate how nonlinear regression succeeds in situations where linear regression fails. The nonlinear regression model differs from the linear regression model in two major aspects: (i) The linear activation function is replaced by a sigmoid function (or a hyperbolic tangent function), and (ii) a logarithmic loss function is used to measure approximation errors instead of a quadratic loss. The resulting optimization problem is not well-posed in the sense that a global minimum cannot be attained at any finite parameter vector. Nevertheless, the gradient descent method can be employed as an efficient numerical algorithm to obtain a good approximate solution. The logic gate examples illustrate how machine learning can achieve accurate and efficient predictions without explicitly solving the open problem of global optimization.

4.1 Nonlinear Data

We use a simple example to illustrate how linear regression fails in fitting nonlinear data. Consider a logic gate with two binary inputs x_1 and x_2 and one binary output y. The dimension is $d = 2$, and the sample size is $n = 4$. Table 4.1 illustrates the input and output data of the AND gate.

For convenience, we use 1 and -1 to denote True and False, respectively. The input matrix of the AND gate can be written as

$$x = \begin{pmatrix} 1 & 1 & 1 \\ 1 & 1 & -1 \\ 1 & -1 & 1 \\ 1 & -1 & -1 \end{pmatrix} \in \mathbb{R}^{4\times 3}. \tag{4.1}$$

X.-S. Wang, C. Wang, *Machine Learning in Data Processing*, Forum for Interdisciplinary Mathematics, https://doi.org/10.1007/978-3-032-20855-2_4

Table 4.1 The AND gate

x_1	x_2	y
True	True	True
True	False	False
False	True	False
False	False	False

The output vector of the AND gate is

$$y = \begin{pmatrix} 1 \\ -1 \\ -1 \\ -1 \end{pmatrix} \in \mathbb{R}^4. \tag{4.2}$$

A simple calculation gives the parameter vector

$$p = (x^T x)^{-1}(x^T y) = \begin{pmatrix} -0.5 \\ 0.5 \\ 0.5 \end{pmatrix} \in \mathbb{R}^3$$

and the prediction vector

$$Y = xp = \begin{pmatrix} 0.5 \\ -0.5 \\ -0.5 \\ -1.5 \end{pmatrix} \in \mathbb{R}^4.$$

Since the linear regression model fails to give a satisfactory prediction, we need to use a nonlinear regression model to fit the data.

4.2 Sigmoid Function

The sigmoid function is usually referred to the following function:

$$S(t) := \frac{1}{1+e^{-t}}, \tag{4.3}$$

which is increasing and bounded by its two horizontal asymptotes 0 and 1. However, from mathematical point of view, any function obtained by scaling and shifting of $S(t)$ will have equivalent effects as $S(t)$ in nonlinear regression and hence should still be called the sigmoid function. In other words, for any real constants α_1, β_1, α_0, and β_0 such that $\alpha_1\beta_1 \neq 0$, the function

$$u(t) := \beta_1 S(\alpha_1 t + \alpha_0) + \beta_0 = \frac{\beta_1}{1 + e^{-\alpha_1 t - \alpha_0}} + \beta_0 \tag{4.4}$$

is also called a sigmoid function. The graph of $u(t)$ is obtained from the graph of $S(t)$ by a sequence of three linear transformations: (i) shift to the left by α_0, (ii) scale by a diagonal matrix $\text{diag}\{1/\alpha_1, \beta_1\}$, and (iii) shift up by β_0. In this book, we shall scale $S(t)$ by a diagonal matrix $\text{diag}\{1/2, 2\}$ and then shift it down by 1; namely, we choose $\alpha_0 = 0, \alpha_1 = \beta_1 = 2$, and $\beta_0 = -1$. The resulting function is the same as the hyperbolic tangent function:

$$u(t) = 2S(2t) - 1 = \frac{2}{1 + e^{-2t}} - 1 = \frac{1 - e^{-2t}}{1 + e^{-2t}} = \frac{e^t - e^{-t}}{e^t + e^{-t}} = \tanh t. \tag{4.5}$$

It is easily seen that $u(t) = \tanh t$ is an odd and increasing function bounded by two horizontal asymptotes ± 1. We also note that

$$u'(t) = \frac{1}{\cosh^2 t} = 1 - \tanh^2 t = 1 - [u(t)]^2. \tag{4.6}$$

The sigmoid function $S(t)$ and its linear transformations are illustrated in Fig. 4.1.

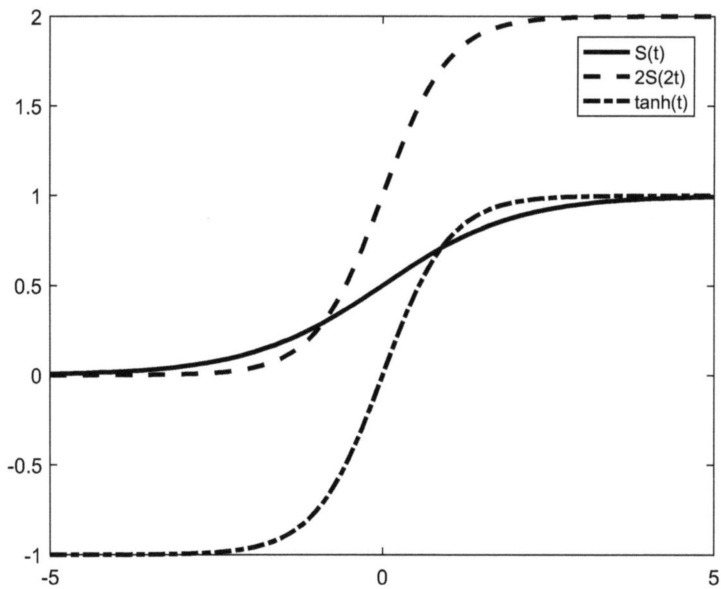

Fig. 4.1 The sigmoid function $S(t) = 1/(1 + e^{-t})$ and its linear transformations

4.3 Optimization Problem

In the nonlinear regression model, the prediction vector is a nonlinear function

$$Y = \tanh(xp), \tag{4.7}$$

where, as usual, $x \in \mathbb{R}^{n\times(d+1)}$ is the input matrix, and $p \in \mathbb{R}^{d+1}$ is the model parameter to be estimated. The above formula is interpreted as $Y_i = \tanh((xp)_i)$ for $i = 1, \cdots, n$. The output vector is binary: $y \in \{-1, 1\}^n$, which is different from that in the linear regression model. Another difference between nonlinear and linear regression model lies in the choice of objective function. In the linear regression model, the Euclidean distance is used to measure the error between the prediction vector and the observation vector. In the nonlinear regression model, we increase the penalty when the prediction is closer to the opposite side of the observation. Given a binary observation $v \in \{-1, 1\}$ and a prediction $u \in (-1, 1)$, we define the cost function

$$C(u, v) := \frac{2\ln 2 - (1+v)\ln(1+u) - (1-v)\ln(1-u)}{2}. \tag{4.8}$$

It is easily seen that

$$\frac{\partial}{\partial u}C(u, v) = -\frac{1+v}{2(1+u)} + \frac{1-v}{2(1-u)} = \frac{u-v}{1-u^2}. \tag{4.9}$$

Note that the value of v is binary: either $v = 1$ or $v = -1$. An equivalent definition of the cost function is

$$C(u, v) = \ln\frac{2}{1+uv} = \begin{cases} \ln[2/(1+u)], & v = 1, \\ \ln[2/(1-u)], & v = -1. \end{cases} \tag{4.10}$$

We consider the following optimization problem.

Optimization Problem
Let $x \in \mathbb{R}^{n\times(d+1)}$ and $y \in \{-1, 1\}^n$ be given. Find $p \in \mathbb{R}^{d+1}$ such that the objective function

$$J(p) := \sum_{i=1}^{n} C(Y_i, y_i) \tag{4.11}$$

is minimized. Here, $Y_i = \tanh((xp)_i)$ for $i = 1, \cdots, n$, and C is the cost function defined in (4.8).

4.4 Gradient Descent Method

Let $J(p)$ be the objective function given in (4.11). We will use the gradient descent method to find a numerical approximation of the parameter vector p that minimizes $J(p)$. For any $i = 1, \cdots, n$, we obtain from

$$Y_i = \tanh((xp)_i) = \tanh\left(\sum_{j=0}^{d} x_{ij} p_j\right)$$

that

$$\frac{\partial Y_i}{\partial p_j} = (1 - Y_i^2)x_{ij}, \quad j = 0, \cdots, d.$$

It then follows from (4.9) and (4.11) that

$$\frac{\partial}{\partial p_j} J(p) = \sum_{i=1}^{n} (Y_i - y_i)x_{ij},$$

for $j = 0, \cdots, d$. The above equation can be written in matrix form as

$$J'(p) = x^T(Y - y). \tag{4.12}$$

The following lemma motivates the gradient descent method.

Lemma 4.1 *Given $p \in \mathbb{R}^{d+1}$ and $J \in C^2(U, \mathbb{R})$, where $U \subset \mathbb{R}^{d+1}$ is an open neighborhood of p. Assume $J'(p) \neq 0$. Then there exists $\alpha_0 > 0$ (which generally depends on p) such that $J(p - \alpha J'(p)) < J(p)$ for any $\alpha \in (0, \alpha_0)$.*

Proof Since $J \in C^2(U, \mathbb{R})$ and $J'(p) \neq 0$, we can find a small $\alpha_0 > 0$ such that $p - \alpha J'(p) \in U$ and

$$\alpha |[J'(p)]^T J''(p - \alpha J'(p)) J'(p)| < \|J'(p)\|_2^2$$

for all $\alpha \in (0, \alpha_0)$. It then follows from Taylor expansion that

$$J(p - \alpha J'(p)) - J(p) = -\alpha \|J'(p)\|_2^2 + \frac{\alpha^2}{2} [J'(p)]^T J''(p - \xi J'(p)) J'(p),$$

for some $\xi \in (0, \alpha)$. Coupling the above two inequalities gives $J(p - \alpha J'(p)) < J(p)$. This completes the proof. □

The gradient descent method is an iterative method to find the minimizer along the direction of descent gradient; namely,

$$p^{(k+1)} = p^{(k)} - \alpha J'(p^{(k)}) = p^{(k)} - \alpha x^T [\tanh(xp^{(k)}) - y], \tag{4.13}$$

where $\alpha > 0$ is called the *l*earning rate, which is a hyperparameter to be tuned in practical applications.

4.5 Python Code

We use the nonlinear regression model to fit the AND gate. The gradient descent method is implemented in the following Python code:

Nonlinear Regression Code

```
import numpy as np

def nonlinear_regression(x,y,al,iter_max):
    d=x.shape[1]-1
    p=np.random.normal(0,1,(3,1))
    for iter in range(iter_max):
        p=p-al*np.dot(x.T,np.tanh(np.dot(x,p))-y)
    return p

x=np.array([[1,1,1],[1,1,-1],[1,-1,1],[1,-1,-1]])
y=np.array([[1],[-1],[-1],[-1]])

p=nonlinear_regression(x,y,al=0.1,iter_max=1000)
print(p)
Y=np.tanh(np.dot(x,p))
print(np.concatenate((Y,y),axis=1))
```

The result of numerical iteration is given as below:

Iteration Result

```
[[-2.99576037]
 [ 2.99576042]
 [ 2.99576041]]
[[ 0.99501275  1.        ]
 [-0.99501275 -1.        ]
```

(continued)

```
 [-0.99501275 -1.        ]
 [-0.99999997 -1.        ]]
```

It is noted that the nonlinear regression model fits the AND gate with a high accuracy.

4.6 Discussions

We first discuss how the nonlinear regression model works for the AND gate. Note from the result of numerical iteration via the gradient descent method in Sect. 4.5 that $p = (-M, M, M)^T \in \mathbb{R}^3$, where $M \approx 3$. A simple calculation gives

$$xp = \begin{pmatrix} 1 & 1 & 1 \\ 1 & 1 & -1 \\ 1 & -1 & 1 \\ 1 & -1 & -1 \end{pmatrix} \begin{pmatrix} -M \\ M \\ M \end{pmatrix} = \begin{pmatrix} M \\ -M \\ -M \\ -3M \end{pmatrix}.$$

Hence, the prediction vector is

$$Y = \tanh(xp) = \begin{pmatrix} \tanh(M) \\ -\tanh(M) \\ -\tanh(M) \\ -\tanh(3M) \end{pmatrix}.$$

Note from Fig. 4.1 that the function $\tanh(t)$ has two horizontal asymptotes ± 1; namely, $\tanh(t) \to \pm 1$ as $t \to \pm\infty$. This implies that $\tanh(M) \approx 1$ for large $M > 0$. Actually, when $M = 3$, we have $\tanh(M) \approx 0.995$ and $\tanh(3M) \approx 0.99999997$. Hence, the prediction vector Y calculated in the above formula can be regarded as a good approximation of the observation vector $y = (1, -1, -1, -1)^T$.

Finding the global minimum of a general objective function, with or without constraints, remains an open problem in optimization. The effectiveness of machine learning lies in the fact that an accurate approximation of the parameter vector need not correspond to a global minimum, which in some cases may not even exist. Instead, there often exist infinitely many parameter vectors that yield sufficiently good approximations, and machine learning algorithms are able to identify such solutions efficiently.

For example, in the nonlinear regression model of the AND gate, the parameter vector $p = (-M, M, M)^T$ is a good solution for any sufficiently large $M > 0$. This illustrates that, in practice, it is neither necessary nor typically possible to determine the global minimum in a machine learning model. Rather, it suffices to perform a finite number of numerical iterations, such as those generated by the gradient descent method, to obtain predictions that closely approximate the observed data.

In the case of the nonlinear regression model for the AND gate, we show in the following proposition that the global minimum of the objective function $J(p)$ in (4.11) cannot be attained at any finite parameter vector.

Proposition 4.1 *Consider the AND gate with input matrix x and output vector y given as in* (4.1) *and* (4.2)*, respectively. Let $J(p)$ be the objective function defined in* (4.11)*. We have $J(p) > 0$ for all $p \in \mathbb{R}^3$. Moreover, we have $J(q(M)) \to 0$ as $M \to \infty$, where $q(M) := (-M, M, M)^T \in \mathbb{R}^3$.*

Proof Note from (4.5) that $C(u, v) > 0$ for any $u \in (-1, 1)$ and $v \in \{-1, 1\}$. Hence, we obtain from (4.11) that $J(p) > 0$ for all $p \in \mathbb{R}^3$. Moreover, a simple calculation gives

$$\tanh(xq(M)) = \begin{pmatrix} \tanh(M) \\ -\tanh(M) \\ -\tanh(M) \\ -\tanh(3M) \end{pmatrix} \to \begin{pmatrix} 1 \\ -1 \\ -1 \\ -1 \end{pmatrix} = y$$

as $M \to \infty$. Thus,

$$\lim_{M\to\infty} J(q(M)) = \sum_{i=1}^{4} C(y_i, y_i) = 0.$$

This completes the proof. □

The above proposition implies that the optimization problem for the nonlinear regression model of the AND gate is not well-posed because the global minimum does not exist. The amazing thing in machine learning is that the gradient descent method can still find a good approximation efficiently even though the global minimum may not exist. The open problem of global optimization is not solved but effectively avoided in machine learning.

Nonlinear regression and linear regression differ in three aspects: the observation vector, the model, and the cost function. However, the derivatives of the objective function and the iterative functions of the gradient descent method for both regression models have the similar formulas. A comparison of these two models is listed in Table 4.2.

Problems

4.1 Use a linear regression model to fit the OR gate. Find the parameter vector and the prediction vector.

4.2 Verify that the function $u(t) = (1 + e^{-2t})^{-1}$ satisfies the differential equation $u'(t) = u(t)[1 - u(t)]$ with initial condition $u(0) = 1/2$.

Table 4.2 The comparison between nonlinear regression and linear regression

	Linear regression	Nonlinear regression
Input matrix	$x \in \mathbb{R}^{n\times(d+1)}$	$x \in \mathbb{R}^{n\times(d+1)}$
Observation vector	$y \in \mathbb{R}^n$	$y \in \{-1, 1\}^n$
Model	$Y = xp \in \mathbb{R}^n$	$Y = \tanh(xp) \in (-1, 1)^n$
Parameter vector	$p \in \mathbb{R}^{d+1}$	$p \in \mathbb{R}^{d+1}$
Cost function	$C(u, v) = (u - v)^2/2$	$C(u, v) = \ln[2/(1 + uv)]$
Objective function	$J(p) = \sum_{i=1}^n C(Y_i, y_i)$	$J(p) = \sum_{i=1}^n C(Y_i, y_i)$
Gradient	$J'(p) = x^T(Y - y)$	$J'(p) = x^T(Y - y)$
Gradient descent method	$p^{(k+1)} = p^{(k)} - \alpha J'(p^{(k)})$	$p^{(k+1)} = p^{(k)} - \alpha J'(p^{(k)})$

4.3 Verify that the function $u(t) = (1 + e^{-2t})^{-1}$ is the cumulative distribution function of the logistic distribution with location parameter 0 and scale parameter $1/2$; namely,

$$u(t) = \int_{-\infty}^{t} v(s)ds,$$

where $v(s) = 2e^{-2s}(1 + e^{-2s})^{-2}$ is the probability density function.

4.4 List the sequence of linear transformations from $u(t) = \tanh t$ to $S(t) = (1 + e^{-t})^{-1}$.

4.5 Let $v \in \{-1, 1\}$, $u \in (-1, 1)$, and $C(u, v)$ be defined in (4.8). Prove (4.10).

4.6 Let $J(p)$ be the objective function given in (4.11). Verify that

$$J''(p) = x^T \operatorname{diag}(1 - Y^2)x + \lambda I,$$

where I is the identity matrix in $\mathbb{R}^{(d+1)\times(d+1)}$ and $\operatorname{diag}(1 - Y^2) \in \mathbb{R}^{n\times n}$ is a diagonal matrix with diagonal terms being $1 - Y_i^2$ for $i = 1, \cdots, n$. Moreover, prove that $J''(p)$ is positive definite.

4.7 Given $p \in \mathbb{R}^{d+1}$ and $J \in C^2(U, \mathbb{R})$, where $U \subset \mathbb{R}^{d+1}$ is an open neighborhood of p, assume $J'(p) \neq 0$, and choose a small $\alpha_0 > 0$ such that U contains the open ball $B_{\alpha_0\|J'(p)\|_2}(p)$ centered at p with radius $\alpha_0\|J'(p)\|_2$. For any $q = p - \alpha J'(p)$ with $\alpha \in (0, \alpha_0)$, prove that

$$J(q) - J(p) = -\alpha\|J'(p)\|_2^2 + \frac{\alpha^2}{2}[J'(p)]^T J''(p - \xi J'(p))J'(p),$$

for some $\xi \in (0, \alpha)$.

4.8 Use the nonlinear regression model to fit the OR gate; namely, find the numerical approximation of p that minimizes the objective function $J(p)$ in (4.11), where the input matrix x is given as in (4.1), and the output vector is $y = (1, 1, 1, -1)^T \in \mathbb{R}^4$.

4.9 Consider the model $Y = \tanh(xp)$, where x is given as in (4.1). Let $M > 0$ be a sufficiently large constant. Determine which logic gate does the model approximate when p is chosen as each of the following seven vectors:

$$\begin{pmatrix} M \\ M \\ M \end{pmatrix}, \begin{pmatrix} M \\ M \\ -M \end{pmatrix}, \begin{pmatrix} M \\ -M \\ M \end{pmatrix}, \begin{pmatrix} M \\ -M \\ -M \end{pmatrix}, \begin{pmatrix} -M \\ M \\ -M \end{pmatrix}, \begin{pmatrix} -M \\ -M \\ M \end{pmatrix}, \begin{pmatrix} -M \\ -M \\ -M \end{pmatrix}.$$

4.10 Given $J \in C^2(\mathbb{R}^{d+1}, \mathbb{R})$ such that $J''(p)$ is positive definite for all $p \in \mathbb{R}^{d+1}$, prove that $p \in \mathbb{R}^{d+1}$ is a global minimum of J if and only if it is a critical point of J; namely, $J(p) < J(q)$ for all $q \in \mathbb{R}^{d+1}$ if and only if $J'(p) = 0$.

Chapter 5
Shallow Neural Network

Abstract The nonlinear model introduced in the previous chapter consists of applying a sigmoid function to a linear combination of the input data. This model can successfully approximate several logic gates, including the AND, OR, NAND, and NOR gates. However, it fails to represent more complex gates such as the XOR gate (as well as the XAND gate). In this chapter, we introduce intermediate variables and employ composite functions to approximate the XOR gate. These composite functions can be interpreted as a network of interconnected neural nodes, and the intermediate variables correspond to the hidden layers of the network.

5.1 Notations

For convenience, we introduce some notations to denote the operations of logic gates. As in the previous chapter, we use 1 and -1 to denote True and False, respectively. We are considering operations on a binary space $B = \{1, -1\}$. For any $x \in B$, the negative (opposite) of x is denoted by $-x$; namely, False is the negative of True and vice versa.

Let $x_1 \in B$ and $x_2 \in B$ be two input variables of a logic gate. The AND gate is the same as the minimum of x_1 and x_2, which is denoted as

$$x_1 \wedge x_2 = \min\{x_1, x_2\} = \begin{cases} 1, & x_1 = x_2 = 1, \\ -1, & (x_1 + 1)(x_2 + 1) = 0. \end{cases} \tag{5.1}$$

The OR gate is the same as the maximum of x_1 and x_2, which is denoted as

$$x_1 \vee x_2 = \max\{x_1, x_2\} = \begin{cases} 1, & (x_1 - 1)(x_2 - 1) = 0, \\ -1, & x_1 = x_2 = -1. \end{cases} \tag{5.2}$$

The XAND gate is the same as the product of x_1 and x_2, which is denoted as

X.-S. Wang, C. Wang, *Machine Learning in Data Processing*, Forum for Interdisciplinary Mathematics, https://doi.org/10.1007/978-3-032-20855-2_5

$$x_1 \cdot x_2 = \begin{cases} 1, & x_1 = x_2, \\ -1, & x_1 \neq x_2. \end{cases} \tag{5.3}$$

The XOR gate is the negative of the XAND gate: $-(x_1 \cdot x_2)$. One can easily verify the following identities:

$$-(x_1 \wedge x_2) = (-x_1) \vee (-x_2), \tag{5.4}$$

$$-(x_1 \vee x_2) = (-x_1) \wedge (-x_2), \tag{5.5}$$

$$-(x_1 \cdot x_2) = (-x_1) \cdot x_2 = x_1 \cdot (-x_2). \tag{5.6}$$

5.2 The XOR Gate

A logic gate has two binary inputs x_1 and x_2 and one binary output y. The corresponding data has a dimension of $d = 2$ and a size of $n = 4$. Table 5.1 illustrates the input and output data of the AND gate, OR gate, NAND gate, NOR gate, and XOR gate, respectively.

The input matrix and the output vector of the XOR gate can be written as

$$x = \begin{pmatrix} 1 & 1 & 1 \\ 1 & 1 & -1 \\ 1 & -1 & 1 \\ 1 & -1 & -1 \end{pmatrix} \in \mathbb{R}^{4\times 3}, \tag{5.7}$$

and

$$y = \begin{pmatrix} -1 \\ 1 \\ 1 \\ -1 \end{pmatrix} \in \mathbb{R}^4, \tag{5.8}$$

respectively. If we run the nonlinear regression code in Sect. 4.5 with the above x and y, the simulation result is given below:

Table 5.1 The logic gates with 1 and − 1 denoting True and False, respectively

x_1	x_2	AND	OR	NAND	NOR	XOR
1	1	1	1	−1	−1	−1
1	−1	−1	1	1	−1	1
−1	1	−1	1	1	−1	1
−1	−1	−1	−1	1	1	−1

Simulation Result

```
[[ 1.08017916e-17]
 [ 7.93636823e-18]
 [-2.85865445e-17]]
[[-9.84838471e-18 -1.00000000e+00]
 [ 4.73247043e-17  1.00000000e+00]
 [-2.57211212e-17  1.00000000e+00]
 [ 3.14519678e-17 -1.00000000e+00]]
```

The parameter vector is close to $(0, 0, 0)^T$, and the prediction vector is also close to the zero vector $(0, 0, 0, 0)^T$, which fails to approximate the observation vector of XOR gate given in (5.8).

5.3 Composite Function

A careful examination of Table 5.1 shows that the XOR gate can be regarded as the AND gate whose inputs are the outputs of the NAND gate and the OR gate; namely, for any $x_1 \in B$ and $x_2 \in B$, we have

$$-x_1 \cdot x_2 = [-(x_1 \wedge x_2)] \wedge (x_1 \vee x_2). \tag{5.9}$$

Since each of the AND, NAND, and OR gates can be approximated by a nonlinear sigmoid function with a linear combination, we approximate y as a composite function of x_1 and x_2 in the sense that y is a function of two intermediate variables z_1 and z_2 that approximates the AND gate, where z_1 and z_2 are functions of x_1 and x_2 that approximate the NAND gate and the OR gate, respectively. This suggests us to propose the following model. For each $i = 1, 2, 3, 4$, we define

$$Y_i = \tanh(z_{i0}p_0 + z_{i1}p_1 + z_{i2}p_2), \tag{5.10}$$

where $z_{i0} = 1$ and

$$z_{i1} = \tanh(x_{i0}q_0 + x_{i1}q_1 + x_{i2}q_2), \tag{5.11}$$

$$z_{i2} = \tanh(x_{i0}r_0 + x_{i1}r_1 + x_{i2}r_2). \tag{5.12}$$

Note that the composite function has a total of nine parameters: p_j, q_j, and r_j with $j = 0, 1, 2$. The objective function is given by

$$J(p, q, r) := \sum_{i=1}^{n} C(Y_i, y_i), \tag{5.13}$$

where C is the cost function defined in (4.8). We also recall from (4.9) that

$$\frac{\partial C}{\partial Y_i} = \frac{Y_i - y_i}{1 - Y_i^2}.$$

We need to calculate the partial derivatives $J_p, J_q, J_r \in \mathbb{R}^3$ so as to implement the gradient descent method. By the chain rule, we obtain the j-th component of J_p as

$$\frac{\partial J}{\partial p_j} = \sum_{i=1}^{n} \frac{\partial C}{\partial Y_i} \frac{\partial Y_i}{\partial p_j} = \sum_{i=1}^{n} (Y_i - y_i) z_{ij}, \tag{5.14}$$

for each $j = 0, 1, 2$. Similarly, we have the j-th components of J_q and J_r, respectively,

$$\frac{\partial J}{\partial q_j} = \sum_{i=1}^{n} \frac{\partial C}{\partial Y_i} \frac{\partial Y_i}{\partial z_{i1}} \frac{\partial z_{i1}}{\partial q_j} = \sum_{i=1}^{n} (Y_i - y_i) p_1 (1 - z_{i1}^2) x_{ij}, \tag{5.15}$$

$$\frac{\partial J}{\partial r_j} = \sum_{i=1}^{n} \frac{\partial C}{\partial Y_i} \frac{\partial Y_i}{\partial z_{i2}} \frac{\partial z_{i2}}{\partial r_j} = \sum_{i=1}^{n} (Y_i - y_i) p_2 (1 - z_{i2}^2) x_{ij}, \tag{5.16}$$

for $j = 0, 1, 2$. In matrix form, we denote the gradient vectors as

$$J_p = (\frac{\partial J}{\partial p_0}, \frac{\partial J}{\partial p_1}, \frac{\partial J}{\partial p_2})^T, \; J_q = (\frac{\partial J}{\partial q_0}, \frac{\partial J}{\partial q_1}, \frac{\partial J}{\partial q_2})^T, \; J_r = (\frac{\partial J}{\partial r_0}, \frac{\partial J}{\partial r_1}, \frac{\partial J}{\partial r_2})^T$$

and obtain

$$J_p = z^T (Y - y), \tag{5.17}$$

$$J_q = x^T [(Y - y) * (1 - z_1 * z_1)] p_1, \tag{5.18}$$

$$J_r = x^T [(Y - y) * (1 - z_2 * z_2)] p_2, \tag{5.19}$$

where $z_1 = (z_{11}, \cdots, z_{41})^T$, $z_2 = (z_{12}, \cdots, z_{42})^T$, and $*$ denotes the component-wise multiplication; namely, if $A \in \mathbb{R}^{n \times m}$ and $B \in R^{n \times m}$, then $A * B \in \mathbb{R}^{n \times m}$ such that $(A * B)_{ij} = A_{ij} B_{ij}$ for each $i = 1, \cdots, n$ and $j = 1, \cdots, m$. We then write the gradient descent method as

$$p^{(k+1)} = p^{(k)} - \alpha J_p^{(k)}, \; q^{(k+1)} = q^{(k)} - \alpha J_q^{(k)}, \; r^{(k+1)} = r^{(k)} - \alpha J_r^{(k)}, \tag{5.20}$$

where $\alpha > 0$ is the learning rate.

5.4 Python Code

The Python code for the gradient descent method is given below:

XOR Code

```
import numpy as np

x=np.array([[1,1,1],[1,1,-1],[1,-1,1],[1,-1,-1]])
y=np.array([[-1],[1],[1],[-1]])

al=0.1
iter_max=1000
p=np.random.normal(0,1,(3,1))
q=np.random.normal(0,1,(3,1))
r=np.random.normal(0,1,(3,1))

for iter in range(iter_max):
    z1=np.tanh(np.dot(x,q))
    z2=np.tanh(np.dot(x,r))
    z=np.concatenate((np.ones((4,1)),z1,z2),axis=1)
    Y=np.tanh(np.dot(z,p))
    Jp=np.dot(z.T,Y-y)
    Jq=np.dot(x.T,(Y-y)*(1-z1*z1))*p[1,0]
    Jr=np.dot(x.T,(Y-y)*(1-z2*z2))*p[2,0]
    p=p-al*Jp
    q=q-al*Jq
    r=r-al*Jr

print(np.concatenate((p,q,r),axis=1))
print(np.concatenate((Y,y),axis=1))
```

5.5 Discussions

A sample of simulation result is given as below:

Simulation Result

```
[[-3.08739711  1.80172906  1.5999076 ]
 [ 3.4102806  -1.95476405  1.77937427]
 [ 3.42289907 -1.94572899  1.77547159]]
[[-0.99478607 -1.        ]
 [ 0.99725619  1.        ]
 [ 0.99726898  1.        ]
 [-0.99469775 -1.        ]]
```

From the simulation result, we have the following observations:

- $p \approx (-3, 3, 3)^T$, which is an approximation of the AND gate: $Y \approx z_1 \wedge z_2$.
- $q \approx (2, -2, -2)^T$, which is an approximation of the NAND gate: $z_1 \approx -(x_1 \wedge x_2)$.
- $r \approx (2, 2, 2)^T$, which is an approximation of the OR gate: $z_2 \approx x_1 \vee x_2$.

This simulation result agrees with the identity (5.9). However, since the decomposition of the XOR gate is not unique, the parameter values obtained by gradient descent method with a random initial guess may differ. For instance, another simulation gives $p \approx (3, 3, 3)^T$, $q \approx (-2, 2, -2)^T$, and $r \approx (-2, -2, 2)^T$, which approximates the following decomposition:

$$- x_1 \cdot x_2 = [x_1 \wedge (-x_2)] \vee [(-x_1) \wedge x_2]. \tag{5.21}$$

This example demonstrates that the optimization problem may lack a finite exact solution and can have multiple approximate solutions. Additionally, the objective function is non-convex. It is important to note that, depending on the initial guess, the gradient descent method may sometimes converge to a critical point that does not minimize the objective function.

Our model (5.10)–(5.12) can be interpreted as a shallow neural network $x \to z \to Y$, where $x \in \mathbb{R}^{4\times 3}$ is the input layer, $Y \in \mathbb{R}^4$ is the output layer, and $z \in \mathbb{R}^{4\times 3}$ is the hidden layer. The map from x to z is characterized by an activation (hyperbolic tangent) function together with linear combinations of x determined by a parameter matrix $(q, r) \in \mathbb{R}^{3\times 2}$, while the map from z to Y is characterized by an activation (hyperbolic tangent) function together with a linear combination of z determined by a parameter vector $p \in \mathbb{R}^3$.

Problems

5.1 Given $x_1, x_2 \in B = \{1, -1\}$, prove that $x_1 \vee x_2 = x_2 \vee x_1$ and $x_1 \wedge x_2 = x_2 \wedge x_1$.

5.2 Given $x_1, x_2, x_3 \in B = \{1, -1\}$, prove that $(x_1 \vee x_2) \vee x_3 = x_1 \vee (x_2 \vee x_3)$ and $(x_1 \wedge x_2) \wedge x_3 = x_1 \wedge (x_2 \wedge x_3)$.

5.3 Given $x_1, x_2, x_3 \in B = \{1, -1\}$, prove that

$$(x_1 \vee x_2) \wedge x_3 = (x_1 \wedge x_3) \vee (x_2 \wedge x_3),$$

and

$$(x_1 \wedge x_2) \vee x_3 = (x_1 \vee x_3) \wedge (x_2 \vee x_3).$$

5.4 Verify (5.9) and (5.21).

5.5 Prove that the zero vectors $p = (0, 0, 0)^T$, $q = (0, 0, 0)^T$, and $r = (0, 0, 0)^T$ remain invariant under the gradient descent iteration (5.20).

5.6 Given

$$A = \begin{pmatrix} 1 & 2 \\ 3 & 4 \end{pmatrix},$$

compute A^2 and $A * A$.

Chapter 6
Deep Neural Network

Abstract We generalize the shallow neural network introduced in the previous chapter to a deep neural network. In mathematical terms, a deep neural network can be viewed as a nested composite function comprising multiple layers of intermediate (matrix-valued) variables and several parameter matrices that define the functions connecting consecutive layers. The gradient of the objective function with respect to each parameter matrix is obtained through a straightforward application of the chain rule in multivariate calculus. This process, which proceeds in the reverse order of the forward computation of the objective function, is known as backward propagation.

6.1 The Model

Recall that the model of shallow neural network (5.10)–(5.12) is rewritten as a chain of maps $x \to z \to Y$, where $x \in \mathbb{R}^{4\times 3}$ is the input layer, $Y \in \mathbb{R}^4$ is the output layer, and $z \in \mathbb{R}^{4\times 3}$ is the hidden layer. Now we use a unified notation to denote a deep neural network

$$x^{(0)} \to x^{(1)} \to \cdots \to x^{(L-1)} \to x^{(L)}, \tag{6.1}$$

where $x^{(0)} = x \in \mathbb{R}^{n\times(d+1)}$ is the input layer and $x^{(L)} = (\mathbf{1}_n, Y) \in \mathbb{R}^{n\times(1+1)}$ with $\mathbf{1}_n := (1, \cdots, 1)^T \in \mathbb{R}^n$ is the output layer, while for each $l = 1, \cdots, L-1$, $x^{(l)} \in \mathbb{R}^{n\times(d_l+1)}$ with $x^{(l)}_{10} = \cdots = x^{(l)}_{n0} = 1$ is the hidden layer. The dimension of the l-th layer (with $l = 0, 1, \cdots, L$) is denoted by d_l; in particular, $d_0 = d$ is the number of input features, and $d_L = 1$ is the dimension of the output layer. The depth of the neural network is denoted by L. The map $x^{(l)} \to x^{(l+1)}$ (with $l = 0, 1, \cdots, L-1$) is defined as

$$x^{(l+1)} = (\mathbf{1}_n, \tanh(x^{(l)} p^{(l)})), \tag{6.2}$$

where $p^{(l)} \in \mathbb{R}^{(d_l+1)\times d_{l+1}}$ is the parameter matrix connecting the l-th layer with the $(l+1)$-th layer. In other words, the above map can be written as

X.-S. Wang, C. Wang, *Machine Learning in Data Processing*, Forum for Interdisciplinary Mathematics, https://doi.org/10.1007/978-3-032-20855-2_6

$$x_{ij_{l+1}}^{(l+1)} = \begin{cases} 1, & j_{l+1} = 0, \\ \tanh(\sum_{j_l=0}^{d_l} x_{ij_l}^{(l)} p_{j_l j_{l+1}}^{(l)}), & j_{l+1} = 1, \cdots, d_{l+1}, \end{cases} \tag{6.3}$$

for each $i = 1, \cdots, n, l = 0, \cdots, L-1$, and $j_{l+1} = 0, 1, \cdots, d_{l+1}$. The objective function is given by

$$J(p^{(0)}, \cdots, p^{(L-1)}) := \sum_{i=1}^{n} C(Y_i, y_i), \tag{6.4}$$

where C is the cost function defined in (4.8).

6.2 Backward Propagation

Similar to the case of shallow neural network, we shall calculate the gradients $\partial J / \partial p^{(l)}$ for each $l = 0, \cdots, L-1$. For convenience, we introduce the following notations:

$$u^{(l)} := x^{(l)} p^{(l)} \in \mathbb{R}^{n \times d_{l+1}}, \tag{6.5}$$

$$v^{(l)} := \frac{\partial J}{\partial u^{(l)}} \in \mathbb{R}^{n \times d_{l+1}}, \tag{6.6}$$

$$q^{(l)} := \frac{\partial J}{\partial p^{(l)}} \in \mathbb{R}^{(d_l+1) \times d_{l+1}}, \tag{6.7}$$

for each $l = 0, \cdots, L-1$. The following formulas are elementary but crucial:

$$\frac{dx_{ij_l}^{(l)}}{du_{ij_l}^{(l-1)}} = 1 - [x_{ij_l}^{(l)}]^2, \quad \frac{\partial u_{ij_{l+1}}^{(l)}}{\partial x_{ij_l}^{(l)}} = p_{j_l j_{l+1}}^{(l)}, \quad \frac{\partial u_{ij_{l+1}}^{(l)}}{\partial p_{j_l j_{l+1}}^{(l)}} = x_{ij_l}^{(l)}, \tag{6.8}$$

for each $i = 1, \cdots, n, l = 0, \cdots, L-1, j_l = 0, \cdots, d_l$, and $j_{l+1} = 1, \cdots, d_{l+1}$. Now, we are ready to state the backward propagation:

$$v_{i1}^{(L-1)} = \frac{\partial J}{\partial x_{i1}^{(L)}} \frac{\partial x_{i1}^{(L)}}{\partial u_{i1}^{(L-1)}} = \frac{x_{i1}^{(L)} - y_i}{1 - [x_{i1}^{(L)}]^2} \{1 - [x_{i1}^{(L)}]^2\} = Y_i - y_i, \tag{6.9}$$

$$v_{ij_l}^{(l-1)} = \sum_{j_{l+1}=1}^{d_{l+1}} \frac{\partial J}{\partial u_{ij_{l+1}}^{(l)}} \frac{\partial u_{ij_{l+1}}^{(l)}}{\partial x_{ij_l}^{(l)}} \frac{dx_{ij_l}^{(l)}}{du_{ij_l}^{(l-1)}} = \sum_{j_{l+1}=1}^{d_{l+1}} v_{ij_{l+1}}^{(l)} p_{j_l j_{l+1}}^{(l)} \{1 - [x_{ij_l}^{(l)}]^2\}, \tag{6.10}$$

Table 6.1 Neural network notations

Notation	Dimension	Remark	Relation
$x^{(l)}$	$n \times (d_l + 1)$	Layer	$x^{(l+1)} = (\mathbf{1}_n, \tanh(u^{(l)}))$
$p^{(l)}$	$(d_l + 1) \times d_{l+1}$	Parameter matrix	$u^{(l)} = x^{(l)} p^{(l)}$
$u^{(l)}$	$n \times d_{l+1}$	Linear combination	$u^{(l)} = x^{(l)} p^{(l)}$
$v^{(l)}$	$n \times d_{l+1}$	Gradient	$v^{(l)} = \partial J / \partial u^{(l)}$
$q^{(l)}$	$(d_l + 1) \times d_{l+1}$	Gradient	$q^{(l)} = \partial J / \partial p^{(l)} = [x^{(l)}]^T v^{(l)}$

for each $i = 1, \cdots, n$, $l = L - 1, \cdots, 1$, and $j_l = 1, \cdots, d_l$. Note that the last formula provides a backward recurrence relation for $v^{(l)}$. Finally, we have

$$q^{(l)}_{j_l j_{l+1}} = \sum_{i=1}^{n} \frac{\partial J}{u^{(l)}_{i j_{l+1}}} \frac{\partial u^{(l)}_{i j_{l+1}}}{\partial p^{(l)}_{j_l j_{l+1}}} = \sum_{i=1}^{n} v^{(l)}_{i j_{l+1}} x^{(l)}_{i j_l}, \tag{6.11}$$

for each $l = 0, \cdots, L - 1$, $j_l = 0, \cdots, d_l$, and $j_{l+1} = 1, \cdots, d_{l+1}$. To write the above formulas in matrix form, we shall introduce several notations. Denote by $*$ the componentwise matrix multiplication. Let $\mathbf{1}_{n \times (d_l+1)} \in \mathbb{R}^{n \times (d_l+1)}$ be the matrix with ones on all entries. For any $A \in \mathbb{R}^{n \times (d_l+1)}$, we denote by $\bar{A}^{(1:n)\times(1:d_l)} \in \mathbb{R}^{n \times d_l}$ the submatrix of A obtained by removing the column $j_l = 0$. We can rewrite the above formulas as

$$v^{(L-1)} = Y - y, \tag{6.12}$$

$$v^{(l-1)} = \overline{\{v^{(l)}[p^{(l)}]^T\} * [\mathbf{1}_{n\times(d_l+1)} - x^{(l)} * x^{(l)}]}^{(1:n)\times(1:d_l)}, \quad l = L - 1, \cdots, 1, \tag{6.13}$$

$$q^{(l)} = [x^{(l)}]^T v^{(l)}, \quad l = 0, \cdots, L - 1. \tag{6.14}$$

The gradient descent method is thus to replace $p^{(l)}$ with $p^{(l)} - \alpha q^{(l)}$ during each iteration, where $\alpha > 0$, as usual, denotes the learning rate. The neural network notations are illustrated in Table 6.1.

6.3 Python Code

The Python code for the (deep) neural network is given below:

Neural Network Code

```
import numpy as np

def neural_network(x,y,hidden_layers,al,iter_max):
    n,m=x.shape
    d=[m-1]+hidden_layers+[1]
    L=len(d)-1
    x=[x]+[None]*L
    v=[None]*L
    p=[None]*L
    for l in range(L):
        p[l]=np.random.normal(0,1,(d[l]+1,d[l+1]))
    for iter in range(iter_max):
        for l in range(L):
            u=np.tanh(np.dot(x[l],p[l]))
            x[l+1]=np.concatenate((np.ones((n,1)),u),axis=1)
        Y=x[L][:,1].reshape(n,1)
        v[L-1]=Y-y
        for l in range(L-1,0,-1):
            v[l-1]=(np.dot(v[l],p[l].T)*(1-x[l]*x[l]))[:,1:]
        for l in range(L):
            p[l]=p[l]-al*np.dot(x[l].T,v[l])
    return Y,p

x=np.array([[1,1,1],[1,1,-1],[1,-1,1],[1,-1,-1]])
y=np.array([[-1],[1],[1],[-1]])
Y,p=neural_network(x,y,hidden_layers=[2],al=0.1,
iter_max=1000)
print(p)
print(np.concatenate((Y,y),axis=1))
```

6.4 Discussions

In the Python code, we use the list data type to address the layers.

1. d is a list of $L+1$ numbers, where $d[l]$ with $l = 0, \cdots, L$ corresponds to the dimension in l-th layer. In particular, $d[0]$ is the dimension of the input layer, and $d[L] = 1$ is the dimension of the output layer.

2. x is a list of $L + 1$ matrices, where $x[l] \in \mathbb{R}^{n\times(d_l+1)}$ with $l = 0, \cdots, L$ corresponds to the data of l-th layer. In particular, $x[0]$ is the input data, and $y[0] = (\mathbf{1}_n, Y)$ is the prediction on the output layer.
3. p is a list of L matrices, where $p[l] \in \mathbb{R}^{(d_l+1)\times d_{l+1}}$ with $l = 0, \cdots, L - 1$ corresponds to the parameter matrix connecting the l-th layer to the $l + 1$-th layer.
4. v is a list of L matrices, where $v[l] \in \mathbb{R}^{n\times d_{l+1}}$ with $l = 0, \cdots, L-1$ corresponds to the gradient of J with respect to u.

The four major steps in the main function are:

1. Initial random guess of parameter matrices:

$$p[l] = np.random.normal(0, 1, (d[l] + 1, d[l + 1]))$$

2. Forward propagation:

$$x[l + 1] = np.concatenate((np.ones((n, 1)), u), axis = 1)$$

3. Backward propagation:

$$v[l - 1] = (np.dot(v[l], p[l].T) * (1 - x[l] * x[l]))[:, 1 :]$$

4. Gradient descent iteration:

$$p[l] = p[l] - al * np.dot(x[l].T, v[l])$$

A sample of simulation result is given as below:

Simulation Result

```
[array([[-1.5815611 ,  1.82134624],
       [ 1.76355366,  1.97592353],
       [-1.75918357, -1.96452545]]), array([[ 3.04703861],
       [ 3.38373003],
       [-3.36820299]])]
[[-0.997053   -1.        ]
 [ 0.99426201  1.        ]
 [ 0.99437418  1.        ]
 [-0.9970351  -1.        ]]
```

With only one hidden layer of dimension 2, the simulation results align with those of the shallow neural network discussed in the previous chapter. To adapt this code for a deeper neural network, we simply need to modify the list *hidden_layer* by increasing the number of hidden layers and configuring them with varying dimensions.

Problems

6.1 Let $C = AB$, where $A \in \mathbb{R}^{n\times m}$, $B \in \mathbb{R}^{m\times r}$, and $C \in \mathbb{R}^{n\times r}$. For each $i = 1, \cdots, n$, $j = 1, \cdots, r$, $k = 1, \cdots, n$, $l = 1, \cdots, m$, $s = 1, \cdots, r$, prove that

$$\frac{\partial C_{ij}}{\partial A_{kl}} = \delta_{ik} B_{lj}, \quad \frac{\partial C_{ij}}{\partial B_{ls}} = \delta_{js} A_{il},$$

where δ_{ik} (resp. δ_{js}) is the Kronecker delta symbol which equals 1 when $i = k$ (respectively, $j = s$) and 0 when $i \neq k$ (respectively, $j \neq s$).

6.2 Verify (6.8).

6.3 Verify (6.9) and (6.10).

6.4 Verify (6.13).

6.5 Change the hidden layers in the Python code, and observe the simulation results.

Chapter 7
Batch Normalization

Abstract This chapter introduces batch normalization in deep neural networks, a technique designed to stabilize and accelerate training by normalizing the intermediate variables at each layer. We extend the deep neural network model from the previous chapter, where each layer consists of a linear combination of the previous layer's outputs followed by an activation function, by adding a normalization step prior to the linear combination. Specifically, for each hidden layer, we standardize the activations using the layer-wise mean and standard deviation. The chapter also presents the backward propagation algorithm adapted to batch normalization, detailing how gradients propagate through the normalization step. Key formulas for derivatives with respect to parameters, activations, and normalized variables are derived, highlighting the dependence of each output on all samples in the batch.

7.1 The Model

As in the previous chapter, we consider a neural network denoted by

$$x^{(0)} \to x^{(1)} \to \cdots \to x^{(L-1)} \to x^{(L)}, \tag{7.1}$$

where $x^{(0)} = x \in \mathbb{R}^{n\times(d_0+1)}$ is the input layer and $x^{(L)} = (\mathbf{1}_n, Y) \in \mathbb{R}^{n\times(d_L+1)}$ (with $d_L = 1$ and $\mathbf{1}_n = (1, \cdots, 1)^T \in \mathbb{R}^n$) is the output layer, while for each $l = 1, \cdots, L-1$, $x^{(l)} \in \mathbb{R}^{n\times(d_l+1)}$ is the hidden layer. The dimension of the l-th layer (with $l = 0, 1, \cdots, L$) is denoted by d_l; in particular, d_0 is the number of input features, and $d_L = 1$ is the dimension of the output layer. The depth of the neural network is denoted by L. The map $x^{(l)} \to x^{(l+1)}$ (with $l = 0, 1, \cdots, L-1$) was defined in the previous chapter as a combination of an activation (generalized sigmoid/hyperbolic tangent) function and a linear combination of $x^{(l)}$ involving a parameter matrix $p^{(l)} \in \mathbb{R}^{d_l+1\times d_{l+1}}$. The key idea of batch normalization is to introduce an additional step of normalization before taking the linear combination. In other words, for each $l = 0, 1, \cdots, L-1$, $i = 1, \cdots, n$, and $j_l = 0, \cdots, d_l$, we define

X.-S. Wang, C. Wang, *Machine Learning in Data Processing*, Forum for Interdisciplinary Mathematics, https://doi.org/10.1007/978-3-032-20855-2_7

$$z_{ij_l}^{(l)} = \begin{cases} 1, & j_l = 0, \\ [x_{ij_l}^{(l)} - \mu_{j_l}^{(l)}]/\sigma_{j_l}^{(l)}, & j_l = 1, \cdots, d_l, \end{cases} \tag{7.2}$$

where

$$\mu_{j_l}^{(l)} = \frac{1}{n}\sum_{i=1}^{n} x_{ij_l}^{(l)} \tag{7.3}$$

is the average and

$$\sigma_{j_l}^{(l)} = \sqrt{\frac{1}{n}\sum_{i=1}^{n}[x_{ij_l}^{(l)} - \mu_{j_l}^{(l)}]^2} \tag{7.4}$$

is the standard deviation. Next, we set

$$u^{(l)} = z^{(l)} p^{(l)} \in \mathbb{R}^{n \times d_{l+1}}, \tag{7.5}$$

$$x^{(l+1)} = (\mathbf{1}_n, \tanh(u^{(l)})), \tag{7.6}$$

where $p^{(l)} \in \mathbb{R}^{(d_l+1)\times d_{l+1}}$ is the parameter matrix connecting the l-th layer with the $(l+1)$-th layer, and $\mathbf{1}_n \in \mathbb{R}^n$ is the vector of 1s. The objective function is still given by

$$J(p^{(0)}, \cdots, p^{(L-1)}) := \sum_{i=1}^{n} C(Y_i, y_i), \tag{7.7}$$

where C is the cost function defined in (4.8).

7.2 Backward Propagation

For convenience, we introduce the following notations:

$$v^{(l)} := \frac{\partial J}{\partial u^{(l)}} \in \mathbb{R}^{n \times d_{l+1}}, \tag{7.8}$$

$$w^{(l)} := \frac{\partial J}{\partial z^{(l)}} \in \mathbb{R}^{n \times (d_l+1)}, \tag{7.9}$$

$$q^{(l)} := \frac{\partial J}{\partial p^{(l)}} \in \mathbb{R}^{(d_l+1) \times d_{l+1}}, \tag{7.10}$$

for each $l = 0, \cdots, L-1$. The major difference introduced by normalization is that $u^{(l)}_{ij_{l+1}}$ not only depends on $u^{(l-1)}_{ij_l}$ but also depends on $u^{(l-1)}_{i'j_l}$ for all $i' = 1, \cdots, n$.

For each $l = 0, \cdots, L-1$, $i = 1, \cdots, n$, $i' = 1, \cdots, n$, and $j_l = 1, \cdots, d_l$, we obtain from (7.2), (7.3), and (7.4) that

$$\frac{\partial \mu^{(l)}_{j_l}}{\partial x^{(l)}_{ij_l}} = \frac{1}{n}, \quad \frac{\partial \sigma^{(l)}_{j_l}}{\partial x^{(l)}_{ij_l}} = \frac{x^{(l)}_{ij_l} - \mu^{(l)}_{j_l}}{n\sigma^{(l)}_{j_l}} = \frac{z^{(l)}_{ij_l}}{n}, \tag{7.11}$$

and

$$\frac{\partial z^{(l)}_{i'j_l}}{\partial x^{(l)}_{ij_l}} = \frac{\delta_{ii'} - 1/n}{\sigma^{(l)}_{j_l}} - \frac{x^{(l)}_{i'j_l} - \mu^{(l)}_{j_l}}{[\sigma^{(l)}_{j_l}]^2} \cdot \frac{z^{(l)}_{ij_l}}{n} = \frac{n\delta_{ii'} - 1 - z^{(l)}_{ij_l} z^{(l)}_{i'j_l}}{n\sigma^{(l)}_{j_l}}, \tag{7.12}$$

where $\delta_{ii'}$ is the Kronecker delta symbol which equals 1 when $i = i'$ and 0 when $i \neq i'$.

Similar to (6.8), we also need the following formulas:

$$\frac{\partial u^{(l)}_{ij_{l+1}}}{\partial z^{(l)}_{ij_l}} = p^{(l)}_{j_l j_{l+1}}, \quad \frac{\partial u^{(l)}_{ij_{l+1}}}{\partial p^{(l)}_{j_l j_{l+1}}} = z^{(l)}_{ij_l}, \tag{7.13}$$

for each $l = 0, \cdots, L-1$, $i = 1, \cdots, n$, $j_l = 0, \cdots, d_l$, and $j_{l+1} = 1, \cdots, d_{l+1}$. For simplicity, we introduce a notation $r^{(l)} \in \mathbb{R}^{n \times d_l}$ such that

$$r^{(l)}_{ij_l} := \frac{dx^{(l)}_{ij_l}}{du^{(l-1)}_{ij_l}} = 1 - [x^{(l)}_{ij_l}]^2, \tag{7.14}$$

for all $l = 0, \cdots, L-1$, $i = 1, \cdots, n$, and $j_l = 1, \cdots, d_l$. We can calculate

$$w^{(l)}_{ij_l} = \sum_{j_{l+1}=1}^{d_{l+1}} \frac{\partial J}{\partial u^{(l)}_{ij_{l+1}}} \frac{\partial u^{(l)}_{ij_{l+1}}}{\partial z^{(l)}_{ij_l}} = \sum_{j_{l+1}=1}^{d_{l+1}} v^{(l)}_{ij_{l+1}} p^{(l)}_{j_l j_{l+1}}, \tag{7.15}$$

for all $l = 0, \cdots, L-1$, $i = 1, \cdots, n$, and $j_l = 0, \cdots, d_l$. Now, we are ready to state the backward propagation:

$$v^{(L-1)}_{i1} = \frac{\partial J}{\partial x^{(L)}_{i1}} \frac{\partial x^{(L)}_{i1}}{\partial u^{(L-1)}_{i1}} = \frac{x^{(L)}_{i1} - y_i}{1 - [x^{(L)}_{i1}]^2} \{1 - [x^{(L)}_{i1}]^2\} = Y_i - y_i, \tag{7.16}$$

$$v^{(l-1)}_{ij_l} = \sum_{i'=1}^{n} \frac{\partial J}{\partial z^{(l)}_{i'j_l}} \frac{\partial z^{(l)}_{i'j_l}}{\partial x^{(l)}_{ij_l}} \frac{dx^{(l)}_{ij_l}}{du^{(l-1)}_{ij_l}} = \sum_{i'=1}^{n} w^{(l)}_{i'j_l} \frac{n\delta_{ii'} - 1 - z^{(l)}_{ij_l} z^{(l)}_{i'j_l}}{n\sigma^{(l)}_{j_l}} r^{(l)}_{ij_l}, \tag{7.17}$$

Table 7.1 Batch normalization notations

Notation	Dimension	Remark	Relation
$x^{(l)}$	$n \times (d_l + 1)$	Layer	$x^{(l+1)} = (\mathbf{1}_n, \tanh(u^{(l)}))$
$\mu^{(l)}$	$1 \times d_l$	Average	$\mu^{(l)}_{j_l} = \frac{1}{n}\sum_{i=1}^n x^{(l)}_{ij_l}$
$\sigma^{(l)}$	$1 \times d_l$	Standard deviation	$\sigma^{(l)}_{j_l} = \sqrt{\frac{1}{n}\sum_{i=1}^n [x^{(l)}_{ij_l} - \mu^{(l)}_{j_l}]^2}$
$z^{(l)}$	$n \times (d_l + 1)$	Normalization	$z^{(l)}_{i0} = 1;\ z^{(l)}_{ij_l} = [x^{(l)}_{ij_l} - \mu^{(l)}_{j_l}]/\sigma^{(l)}_{j_l},\ j_l > 0$
$p^{(l)}$	$(d_l + 1) \times d_{l+1}$	Parameter matrix	$u^{(l)} = z^{(l)} p^{(l)}$
$u^{(l)}$	$n \times d_{l+1}$	Linear combination	$u^{(l)} = z^{(l)} p^{(l)}$
$r^{(l)}$	$n \times d_{l+1}$	Derivative of activation	$u^{(l)}_{ij_l} = 1 - [x^{(l)}_{ij_l}]^2,\ j_l > 0$
$v^{(l)}$	$n \times d_{l+1}$	Gradient	$v^{(l)} = \partial J/\partial u^{(l)}$
$w^{(l)}$	$n \times (d_l + 1)$	Gradient	$w^{(l)} = \partial J/\partial z^{(l)} = v^{(l)}[p^{(l)}]^T$
$q^{(l)}$	$(d_l + 1) \times d_{l+1}$	Gradient	$q^{(l)} = \partial J/\partial p^{(l)} = [z^{(l)}]^T v^{(l)}$

for each $i = 1, \cdots, n, l = L - 1, \cdots, 1$, and $j_l = 1, \cdots, d_l$. On account of (7.15), the above formula provides a backward recurrence relation for $v^{(l)}$. Finally, we have

$$q^{(l)}_{j_l j_{l+1}} = \sum_{i=1}^{n} \frac{\partial J}{u^{(l)}_{ij_{l+1}}} \frac{\partial u^{(l)}_{ij_{l+1}}}{\partial p^{(l)}_{j_l j_{l+1}}} = \sum_{i=1}^{n} v^{(l)}_{ij_{l+1}} z^{(l)}_{ij_l}, \tag{7.18}$$

for each $l = 0, \cdots, L - 1$, $j_l = 0, \cdots, d_l$, and $j_{l+1} = 1, \cdots, d_{l+1}$.

The batch normalization notations are illustrated in Table 7.1.

7.3 Python Code

The Python code for the batch normalization is given below:

Batch Normalization Code

```
import numpy as np

def append(A):
 return np.concatenate((np.ones((A.shape[0],1)),A),axis=1)
def cut(A):
 return A[:,1:]
def ave(A):
 n=A.shape[0]
```

(continued)

```
    return np.dot(np.ones((n,n)),A)/n

def batch_normalization(x,y,hidden_layers,al,iter_max):
    n=x.shape[0]
    d=[x.shape[1]-1]+hidden_layers+[1]
    L=len(d)-1
    p,v,z,mu,s=[None]*L,[None]*L,[None]*L,[None]*L,[None]*L
    x=[x]+[None]*L
    for l in range(L):
        p[l]=np.random.normal(0,1,(d[l]+1,d[l+1]))
    for iter in range(iter_max):
        for l in range(L):
            mu[l]=sum(cut(x[l]))/n
            s[l]=np.sqrt(sum((cut(x[l])-mu[l])**2)/n)
            z[l]=append((cut(x[l])-mu[l])/s[l])
            x[l+1]=append(np.tanh(np.dot(z[l],p[l])))
        Y=x[L][:,1].reshape(n,1)
        v[L-1]=Y-y
        for l in range(L-1,0,-1):
            w=np.dot(v[l],p[l].T)
            r=cut(1-x[l]*x[l])
            v[l-1]=cut(w-ave(w)-z[l]*ave(z[l]*w))*r/s[l]
        for l in range(L):
            p[l]=p[l]-al*np.dot(z[l].T,v[l])
    return Y,p,mu,s

x=np.array([[1,1,1],[1,1,-1],[1,-1,1],[1,-1,-1]])
y=np.array([[1],[-1],[-1],[1]])
hidden_layers=[2]
Y,p,mu,s=batch_normalization(x,y,hidden_layers,0.1,1000)
print(np.concatenate((Y,y),axis=1))
```

7.4 Discussions

In the Python code, we have introduced three auxiliary functions to simplify the expressions in the main function:

1. The *append* function is defined as appending a column of ones to the left of a given matrix. For instance, given an input matrix $A \in \mathbb{R}^{n\times d}$, the output of *append* function is the matrix $(\mathbf{1}_n, A) \in \mathbb{R}^{n\times(d+1)}$.

2. The *cut* function is defined as deleting the most left column of a given matrix. It is clear that $cut(append(A)) = A$, but the equation $append(cut(A)) = A$ is not valid unless the first column of A is a column of ones.
3. The *ave* function is defined as a matrix of identical rows with each row being the column average of a given matrix. For instance, given $A \in \mathbb{R}^{n\times d}$, the output of *ave* function is the matrix $(\mathbf{1}_n, \cdots, \mathbf{1}_n)A \in \mathbb{R}^{n\times d}$.

Problems

7.1 Given $\mu = \frac{1}{n}\sum_{i=1}^{n} x_i$ and $\sigma = \sqrt{\frac{1}{n}\sum_{i=1}^{n}(x_i - \mu)^2}$, prove that

$$\frac{\partial \mu}{\partial x_i} = \frac{1}{n}, \quad \frac{\partial \sigma}{\partial x_i} = \frac{x_i - \mu}{n\sigma}.$$

7.2 Given $\mu = \frac{1}{n}\sum_{i=1}^{n} x_i$, $\sigma = \sqrt{\frac{1}{n}\sum_{i=1}^{n}(x_i - \mu)^2}$, and $z_i = (x_i - \mu)/\sigma$, prove that

$$\frac{\partial z_{i'}}{\partial x_i} = \frac{n\delta_{ii'} - 1 - z_i z_{i'}}{n\sigma},$$

where

$$\delta_{ii'} = \begin{cases} 1, & i = i', \\ 0, & i \neq i'. \end{cases}$$

7.3 Verify (7.12).

7.4 Verify (7.13).

7.5 Verify (7.16) and (7.17).

Chapter 8
Support Vector Machine

Abstract Given two finite sets of real numbers, S_+ and S_-, where all numbers in S_+ are positive and all numbers in S_- are negative, one can always select a threshold value (e.g., 0) that separates the two sets, placing all negative numbers to its left and all positive numbers to its right. Among infinitely many such thresholds, there exists an optimal threshold that maximizes the minimal distance from all points in $S_\pm$ to the threshold. This optimal threshold forms the foundation of the support vector machine (SVM). In this chapter, we extend this idea to higher dimensional spaces, $\mathbb{R}^d$, and discuss two equivalent optimization formulations for SVMs.

8.1 Notations

Let $x \in \mathbb{R}^{n\times(d+1)}$ be the input feature matrix of size n and dimension d, where $x_{i0} = 1$ for $i = 1, \cdots, n$. Assume that the observation vector is binary $y \in \{1, -1\}^n$ such that $y_i \in \{1, -1\}$. For each $X_i = (x_{i0}, \cdots, x_{id})^T \in \mathbb{R}^{d+1}$, we denote $\hat{X}_i = (x_{i1}, \cdots, x_{id})^T \in \mathbb{R}^d$. Let $p = (p_0, \cdots, p_d)^T \in \mathbb{R}^{d+1}$ be the parameter vector. Similarly, we denote $\hat{p} = (p_1, \cdots, p_d)^T \in \mathbb{R}^d$. For any vector $u \in \mathbb{R}^d$, the norm is denoted as $\|u\| = \sqrt{u^T u}$. Given any number $t \in \mathbb{R}$, we denote

$$t_+ := \max\{t, 0\} = \begin{cases} t, \ t \geq 0, \\ 0, \ t \leq 0. \end{cases} \tag{8.1}$$

Given a parameter vector p, the output vector is $Y = xp \in \mathbb{R}^n$ such that

$$Y_i = (xp)_i = \sum_{j=0}^{d} x_{ij} p_j = X_i^T p = p_0 + \hat{X}_i^T \hat{p}.$$

The *feasible set* of parameter vectors in $\mathbb{R}^{d+1}$ is defined as

$$D := \{p \in \mathbb{R}^{d+1} : \ y_i(xp)_i \geq 1, \ i = 1, \cdots, n\}. \tag{8.2}$$

X.-S. Wang, C. Wang, *Machine Learning in Data Processing*, Forum for Interdisciplinary Mathematics, https://doi.org/10.1007/978-3-032-20855-2_8

A parameter vector $p \in D$ is called *feasible* in the sense that $y_i(xp)_i \geq 1$ for each $i = 1, \cdots, n$. It is easily seen that if $p \in D$, so is cp for any $c \geq 1$. Throughout this chapter, we assume that the given data $\{x, y\}$ is *separable*; namely, the feasible set D is non-empty. We also assume that the data is *nontrivial*: There exist i and j such that $y_i \neq y_j$. The *normalized feasible set* is defined as

$$E := \{p/\|\hat{p}\| : \ p \in D\}. \tag{8.3}$$

In other words, E is the set of *normalized feasible* parameter vectors such that $\|\hat{p}\| = 1$ and $y_i(xp)_i > 0$ for all $i = 1, \cdot, n$.

Any parameter vector $p \in \mathbb{R}^{d+1}$ with $\|\hat{p}\| > 0$ defines a hyperplane in $\mathbb{R}^d$:

$$H(p) := \{u \in \mathbb{R}^d : \ p_0 + u^T \hat{p} = 0\}. \tag{8.4}$$

The distance from $u \in \mathbb{R}^d$ to the hyperplane $H(p)$ is given by

$$d(u, H(p)) := \min_{v \in H(p)} \|u - v\| = \frac{|p_0 + u^T \hat{p}|}{\|\hat{p}\|}. \tag{8.5}$$

8.2 Optimization Problems

Given a nontrivial separable data $x \in \mathbb{R}^{n \times (d+1)}$ and $y \in \{1, -1\}^n$, we minimize the objective function

$$J(p) := \sum_{i=1}^{n} C(Y_i, y_i) + \frac{\lambda}{2} \sum_{j=1}^{d} p_j^2, \tag{8.6}$$

with cost function

$$C(Y_i, y_i) := \frac{(1 + y_i)(1 - Y_i)_+ + (1 - y_i)(1 + Y_i)_+}{2} = (1 - y_i Y_i)_+, \tag{8.7}$$

where $p \in \mathbb{R}^{d+1}$ is the parameter vectors, $Y = xp \in \mathbb{R}^n$ is the output vector, and $\lambda > 0$ is the regularization parameter. Taking partial derivatives gives

$$\frac{\partial J}{\partial Y_i} = \begin{cases} -y_i, & y_i Y_i < 1, \\ 0, & y_i Y_i > 1, \end{cases} \tag{8.8}$$

and

$$\frac{\partial J}{\partial p_j} = - \sum_{y_i Y_i < 1} y_i x_{ij} + \lambda(1 - \delta_{j0}) p_j, \tag{8.9}$$

where δ_{j0} is the Kronecker delta symbol that equals 1 when $j = 0$ and 0 when $j \neq 0$. The gradient descent method is given by

$$p_j^{(k+1)} = [1 - \alpha\lambda(1 - \delta_{j0})]p_j^{(k)} + \alpha \sum_{y_i Y_i < 1} y_i x_{ij}, \tag{8.10}$$

where $\alpha > 0$ is the learning rate.

Since the data is separable, namely, the feasible set D in (8.2) is non-empty, we have $J(p) = \lambda\|\hat{p}\|^2/2$ for all $p \in D$.

As $\lambda \to 0^+$, the optimization problem of minimizing $J(p)$ converges to the following constrained optimization problem.

Constrained Optimization Problem I
Let $x \in \mathbb{R}^{n\times(d+1)}$ and $y \in \mathbb{R}^n$ be a given nontrivial separable data. Find $p \in D \subset \mathbb{R}^{d+1}$ such that the objective function

$$J_1(p) := \|\hat{p}\| = \sqrt{\sum_{j=1}^{d} p_j^2} \tag{8.11}$$

is minimized.

Proposition 8.1 *Assume that the data $\{x, y\}$ is nontrivial and separable. The objective function $J_1(p)$ in* (8.11) *possesses at least one minimizer, denoted by p^*, in D.*

Proof Since the data is separable, the feasible set D is non-empty. Choose any $q \in D$, and denote

$$r := \sqrt{\sum_{j=1}^{d} q_j^2}.$$

Since the data is nontrivial, we have $r > 0$. Moreover, there exist i_1 and i_2 such that $y_{i_1} = 1$ and $y_{i_2} = -1$. For any $p \in D$ with $\|\hat{p}\| \leq r$, we have $p_0 + \hat{X}_{i_1}^T\hat{p} \geq 1$ and $p_0 + \hat{X}_{i_2}^T\hat{p} \leq -1$. Consequently,

$$p_0 \geq 1 - \hat{X}_{i_1}^T\hat{p} \geq 1 - r\|\hat{X}_{i_1}\|,$$

and

$$p_0 \le -1 - \hat{X}_{i_2}^T \hat{p} \le -1 + r\|\hat{X}_{i_2}\|.$$

This implies that the subset

$$D_r := \{p \in D : \ \|\hat{p}\| \le r\}$$

is bounded. It is easily seen that D_r is also closed. Hence, the objective function J_1 possesses a minimizer, denoted by p^*, in D_r, with $J_1(p^*) \le r$. Since $J_1(p) > r$ for all $p \in D \setminus D_r$, we have $J_1(p^*) \le J_1(p)$ for all $p \in D$. This completes the proof. □

Lemma 8.1 *For any $p \in D$, we have*

$$J_1(p) \ge [\min_i d(\hat{X}_i, H(p/\|\hat{p}\|))]^{-1}. \tag{8.12}$$

Proof For any $p \in D$ and $i = 1, \cdots, n$, we have $y_i(p_0 + \hat{X}_i^T \hat{p}) \ge 1$. Since the data is nontrivial, we also have $\|p\| > 0$. Consequently, $|p_0 + \hat{X}_i^T \hat{p}| = y_i(p_0 + \hat{X}_i^T \hat{p}) \ge 1$ and

$$d(\hat{X}_i, H(p)) = \frac{|p_0 + \hat{X}_i^T \hat{p}|}{\|\hat{p}\|} \ge \frac{1}{J_1(p)}.$$

Since $H(p) = H(p/\|\hat{p}\|)$, the inequality (8.12) follows. □

Lemma 8.1 suggests us to consider the following constrained optimization problem.

Constrained Optimization Problem II

Let $x \in \mathbb{R}^{n\times(d+1)}$ and $y \in \mathbb{R}^n$ be a given nontrivial separable data. Find $p \in E \subset \mathbb{R}^{d+1}$ such that the objective function

$$J_2(p) := \min_i |p_0 + \hat{X}_i^T \hat{p}| \tag{8.13}$$

is maximized.

Proposition 8.2 *Assume that the data $\{x, y\}$ is nontrivial and separable. The objective function $J_2(p)$ in* (8.13) *possesses at least one minimizer, denoted by p^{**}, in E.*

Proof The conclusion follows from the fact that E is compact and J_2 is continuous. □

8.3 Equivalence Theorem

We will prove that the two constrained optimization problems are equivalent.

Theorem 8.1 *The constrained optimization problem I has a unique minimizer $p^* \in D$, and the constrained optimization problem II has a unique maximizer $p^{**} \in E$. Moreover, we have*

$$p^{**} = \frac{p^*}{\|\hat{p}^*\|}, \quad J_1(p^*) = \frac{1}{J_2(p^{**})}. \tag{8.14}$$

Proof On account of Proposition 8.1 and Proposition 8.2, there exist $p^* \in D$ and $p^{**} \in E$ such that $J_1(p^*) \le J_1(p)$ for all $p \in D$ and $J_2(p^{**}) \ge J_2(p)$ for all $p \in E$. For any $p \in E$, we have $y_i(p_0 + \hat{X}_i^T \hat{p}) > 0$ for all $i = 1, \cdots, n$. Hence, we obtain

$$J_2(p) = \min\{J_2^+(p), J_2^-(p)\}, \tag{8.15}$$

where

$$J_2^+(p) = \min_{y_i=1}(p_0 + \hat{X}_i^T \hat{p}) = p_0 + \min_{y_i=1}(\hat{X}_i^T \hat{p}), \tag{8.16}$$

$$J_2^-(p) = \min_{y_i=-1}(-p_0 - \hat{X}_i^T \hat{p}) = -p_0 + \min_{y_i=-1}(-\hat{X}_i^T \hat{p}). \tag{8.17}$$

Note that $J_2(p^{**}) \ge J_2(p)$ for all $p \in E$.

We claim $J_2^+(p^{**}) = J_2^-(p^{**})$; namely, $p_0^{**} = g(\hat{p}^{**})$ with

$$g(u) := \frac{1}{2}[\min_{y_i=-1}(-\hat{X}_i^T u) - \min_{y_i=1}(\hat{X}_i^T u)], \quad u \in \mathbb{R}^d.$$

Otherwise, we may choose $p \in E$ with $\hat{p} = \hat{p}^{**}$ and $p_0 = g(\hat{p})$ to obtain

$$J_2(p) = J_2^+(p) = J_2^-(p) = \frac{J_2^+(p) + J_2^-(p)}{2} = \frac{J_2^+(p^{**}) + J_2^-(p^{**})}{2} > J_2(p^{**}),$$

a contradiction.

Now, we define a function $f \in C(\mathbb{R}^d, \mathbb{R})$ as

$$f(u) := \frac{1}{2}[\min_{y_i=-1}(-\hat{X}_i^T u) + \min_{y_i=1}(\hat{X}_i^T u)], \tag{8.18}$$

for $u \in \mathbb{R}^d$. It follows from the above argument that $J_2(p^{**}) = f(\hat{p}^{**})$. Moreover, by definition, we observe that

$$f(cu) = cf(u), \quad f(u+v) \geq f(u) + f(v), \tag{8.19}$$

for any $c > 0$ and $u, v \in \mathbb{R}^d$. Since $\|\hat{p}\| = 1$ and $f(\hat{p}) > 0$ for $p \in E$, the function $f(u)$ has a unique maximizer on the unit sphere $S := \{u \in \mathbb{R}^d, \ \|u\| = 1\}$.

We show that the maximizer of f on S is exactly $\hat{p}^{**}$. This is because for any $u \in S$ with $f(u) > 0$, we may choose $p \in \mathbb{R}^{d+1}$ with $\hat{p} = u$ and $p_0 = g(u)$ such that $p \in E$ and $J_2(p) = J_2^+(p) = J_2^-(p) = f(\hat{p})$, and hence $f(\hat{p}^{**}) = J_2(p^{**}) \geq J_2(p) = f(\hat{p}) = f(u)$. On the other hand, if $u \in S$ with $f(u) \leq 0$, then $f(\hat{p}^{**}) > 0 \geq f(u)$. This proves that $\hat{p}^{**}$ is the unique maximizer of f on S. In view of $p_0^{**} = g(\hat{p}^{**})$, the maximizer of J_2 is also unique.

Finally, since $p^*/J_1(p^*) \in E$ and $p^{**}/J_2(p^{**}) \in D$, we have

$$J_2(p^{**}) \geq J_2(p^*/J_1(p^*)) \geq \frac{1}{J_1(p^*)} \geq \frac{1}{J_1(p^{**}/J_2(p^{**}))} = J_2(p^{**}).$$

The above inequalities should be equalities, and the identities (8.14) are proved. Finally, the uniqueness of p^* follows from the uniqueness of p^{**}. □

Problems

8.1 Given $p \in \mathbb{R}^{d+1}$ with $\|\hat{p}\| > 0$. Let $H(p)$ be the hyperplane defined in (8.4). For any $u \in \mathbb{R}^d$, prove that

$$\min_{v \in H(p)} \|u - v\| = \frac{|p_0 + u^T \hat{p}|}{\|\hat{p}\|}.$$

8.2 Verify that $H(p) = H(cp)$ for any $c \neq 0$ and $p \in \mathbb{R}^{d+1}$ with $\|p\| > 0$.

8.3 Given $Y_i \in \mathbb{R}$ and $y_i \in \{1, -1\}$, prove that

$$\frac{(1+y_i)(1-Y_i)_+ + (1-y_i)(1+Y_i)_+}{2} = (1 - y_i Y_i)_+.$$

8.4 Assume that D in (8.2) is non-empty. Prove that $J(p) = \lambda \|\hat{p}\|^2/2$ for all $p \in D$.

8.5 Prove that the data $\{x, y\}$ is nontrivial if and only if $\|\hat{p}\| > 0$ for any $p \in D$.

8.6 Assume that the data $\{x, y\}$ is nontrivial and separable. Prove that the feasible set D defined in (8.2) is closed in $\mathbb{R}^{d+1}$.

8.7 Assume that the data $\{x, y\}$ is nontrivial and separable. Prove that the normalized feasible set E defined in (8.3) is compact (i.e., bounded and closed) in $\mathbb{R}^{d+1}$.

8.8 Show that $p/J_2(p) \in D$ if $p \in E$.

8.9 Prove that E defined in (8.3) is the same as

$$E = \{p \in \mathbb{R}^{d+1} : \ \|\hat{p}\| = 1, \ y_i(xp)_i > 0, \ i = 1, \cdots, n\}.$$

8.10 Let E and f be defined as in (8.3) and (8.18), respectively. Prove that $f(\hat{p}) > 0$ for any $p \in E$.

8.11 Verify (8.19).

8.12 Assume $h \in C(\mathbb{R}^d, \mathbb{R})$ is convex and $h(cu) = ch(u)$ for any $c > 0$ and $u \in \mathbb{R}^d$. If the minimum of h on the unit sphere $S = \{u \in \mathbb{R}^d : \|u\| = 1\}$ is negative, prove that the minimizer on S is unique.

Chapter 9
Gradient Methods

Abstract This chapter introduces several fundamental gradient-based iterative methods for solving linear systems and optimization problems, including the conjugate gradient method, the Jacobi method, and the heavy ball method. We begin with the conjugate gradient method for symmetric positive definite matrices, presenting the derivation of the iteration formulas, orthogonality properties, and the underlying vector space structure. The method is shown to converge to the exact solution in at most m steps for an m-dimensional system. Next, we discuss one-step iterative methods, including the Jacobi and gradient descent methods, and analyze their convergence using eigenvalue techniques. The concept of a learning rate is introduced, and we formulate a minimax optimization problem to determine the optimal learning rate for the fastest convergence. Finally, the chapter extends these ideas to two-step methods, or heavy ball methods, which incorporate a momentum term to accelerate convergence. We analyze the spectral radius of the associated iteration matrix and present a minimax problem to optimize both the learning rate and momentum parameter.

9.1 Conjugate Gradient Method

Let $A \in \mathbb{R}^{m\times m}$ be a symmetric and positive definite matrix. We may define an inner product on $\mathbb{R}^m$ as

$$(u, w) = u^T Aw, \quad u, w \in \mathbb{R}^m. \tag{9.1}$$

Given $b \in \mathbb{R}^m$, we intend to solve the linear system

$$Av = b. \tag{9.2}$$

The above equation is a generalization of (2.12) in linear regression. In general, the computation cost of $v = A^{-1}b$ is $O(m^3)$. We shall introduce an iteration method to find an approximate solution using less computation time.

X.-S. Wang, C. Wang, *Machine Learning in Data Processing*, Forum for Interdisciplinary Mathematics, https://doi.org/10.1007/978-3-032-20855-2_9

Let $v_1 \in \mathbb{R}^m$ be any initial guess of solution, and we denote $e_1 = v_1 - v$ and $r_1 = Av_1 - b = Ae_1$ to be the error vector and residual vector, respectively. We also initialize $u_1 = r_1$. If $u_1 = 0$, then we have found the solution. Otherwise, we define $r_2 = Ae_2$ with

$$e_2 = e_1 - \alpha_1 u_1, \quad \alpha_1 = \frac{(e_1, u_1)}{(u_1, u_1)}.$$

Note that $\alpha_1 u_1$ is the projection of e_1 on u_1 and e_2 is perpendicular to u_1; namely, $(e_2, u_1) = 0$. We also define

$$u_2 = r_2 - \frac{(r_2, u_1)}{(u_1, u_1)} u_1$$

such that $(u_2, u_1) = 0$.

Now, we assume that e_j, $r_j = Ae_j$ and u_j with $j = 1, \cdots, k$ have been calculated. If $u_k = 0$, then we are done. Otherwise, we define

$$r_{k+1} = Ae_{k+1}, \tag{9.3}$$

with

$$e_{k+1} = e_k - \alpha_k u_k, \quad \alpha_k = \frac{(e_k, u_k)}{(u_k, u_k)}. \tag{9.4}$$

We also use the Gram-Schmidt process to define

$$u_{k+1} = r_{k+1} - \sum_{j=1}^{k} \frac{(r_k, u_j)}{(u_j, u_j)} u_j \tag{9.5}$$

such that

$$(u_k, u_j) = 0, \quad j < k. \tag{9.6}$$

We have the following proposition.

Proposition 9.1 *The following four vector spaces are identical:*

$$\begin{aligned} S_k^{(1)} &:= \mathrm{span}\{r_1, r_2, \cdots, r_k\}, \\ S_k^{(2)} &:= \mathrm{span}\{u_1, u_2, \cdots, u_k\}, \\ S_k^{(3)} &:= \mathrm{span}\{r_1, Ar_1, \cdots, Ar_{k-1}\}, \\ S_k^{(4)} &:= \mathrm{span}\{u_1, Au_1, \cdots, Au_{k-1}\}. \end{aligned}$$

Proof We prove the statement by induction. First, since $u_1 = r_1$, we have

$$S_k^{(1)} = S_k^{(2)} = S_k^{(3)} = S_k^{(4)}$$

for $k = 1$. Now, we assume that the above identities are true for all $k = 1, \cdots, K$. In view of (9.5), we have $S_k^{(1)} = S_k^{(2)}$ and $S_k^{(3)} = S_k^{(4)}$ for $k = K + 1$. Moreover, since

$$r_{k+1} - r_k = A(e_{k+1} - e_k) = -\alpha_k A u_k$$

by (9.4), we also have $S_k^{(1)} = S_k^{(4)}$ for $k = K + 1$. Therefore, the four vector spaces are identical for $k = K + 1$. This completes the proof. □

On account of Proposition 9.1, we denote

$$\begin{aligned} S_k :=& \operatorname{span}\{r_1, r_2, \cdots, r_k\} = \operatorname{span}\{u_1, u_2, \cdots, u_k\} \\ =& \operatorname{span}\{r_1, Ar_1, \cdots, Ar_{k-1}\} = \operatorname{span}\{u_1, Au_1, \cdots, Au_{k-1}\}. \end{aligned} \tag{9.7}$$

We have the following corollary.

Corollary 9.1 *For any $j < k$, we have*

$$r_k^T u_j = (e_k, u_j) = 0, \tag{9.8}$$

$$(r_{k+1}, u_j) = (e_{k+1}, Au_j) = 0. \tag{9.9}$$

Proof If $k = 2$, then (9.4) implies that $r_2^T u_1 = (e_2, u_1) = 0$. Now, we assume that (9.8) holds for any $1 \le j < k \le K$. By (9.6) and (9.4), we have $(e_{k+1}, u_k) = 0$ and

$$(e_{k+1}, u_j) = (e_k, u_j) - \alpha_k (u_k, u_j) = 0$$

for $1 \le j < k = K$. Thus, (9.8) is valid for all $1 \le j < k = K + 1$. By induction, (9.8) is proved for all $1 \le j < k$. Hence, e_{k+1} is perpendicular to the space S_k defined in (9.7), which yields (9.9). □

In view of (9.9), we can rewrite (9.5) as

$$u_{k+1} = r_{k+1} + \beta_k u_k, \tag{9.10}$$

where

$$\beta_k := -\frac{(r_{k+1}, u_k)}{(u_k, u_k)} = \frac{r_{k+1}^T (r_{k+1} - r_k)}{u_k^T (r_k - r_{k+1})} = \frac{r_{k+1}^T r_{k+1}}{r_k^T r_k}. \tag{9.11}$$

The projection (9.4) can be rewritten as

$$v_{k+1} = v_k - \alpha_k u_k, \tag{9.12}$$

where

$$\alpha_k = \frac{(e_k, u_k)}{(u_k, u_k)} = \frac{r_k^T r_k}{u_k^T A u_k}. \tag{9.13}$$

Conjugate Gradient Method Pseudo Code
Input: $A \in \mathbb{R}^{m\times m}$, $b \in \mathbb{R}^m$
Initialize: $v_1 = 0 \in \mathbb{R}^m$, $r_1 = Av_1 - b$, $u_1 = r_1$
Loop: for $k = 1, 2, \cdots$, while $r_k^T r_k > \varepsilon$, compute the following two lines

- Projection: $v_{k+1} = v_k - \alpha_k u_k$, $r_{k+1} = r_k - \alpha_k A u_k$ where $\alpha_k = r_k^T r_k/(u_k^T A u_k)$
- Orthogonalization: $u_{k+1} = r_{k+1} + \beta_k u_k$ where $\beta_k = r_{k+1}^T r_{k+1}/(r_k^T r_k)$

Output: v_k

Theorem 9.1 *If $u_1, \cdots, u_m$ are nonzero vectors, then $r_{m+1} = 0$, $e_{m+1} = 0$, and $v_{m+1} = v$.*

Proof Since $(u_k, u_j) = 0$ for $j < k$, the nonzero vectors $u_1, \cdots, u_m$ are orthogonal to each other. Hence, $S_m = \text{span}\{u_1, \cdots, u_m\} = \mathbb{R}^m$ and $u_1, \cdots, u_m$ are orthogonal basis in $\mathbb{R}^m$. By (9.9), we have $(r_{m+1}, u_j) = 0$ for all $j = 1, \cdots, m$. Consequently, $r_{m+1} = 0$, which implies $e_{m+1} = A^{-1} r_{m+1} = 0$ and $v_{m+1} = v + e_{m+1} = v$. This completes the proof. □

9.2 One-Step Method

Let $A \in \mathbb{R}^{m\times m}$ be a nonsingular matrix. Choose a diagonal matrix $D \in \mathbb{R}^{m\times m}$ with positive diagonal entries such that all eigenvalues of $D^{-1}A$ are positive. The linear system $Av = b$ can be rewritten as $Dv = Dv + \alpha(b - Av)$ for any $\alpha > 0$. Multiplying by D^{-1} on both sides gives

$$v = D^{-1}[Dv + \alpha(b - Av)] = v - \alpha D^{-1}(Av - b).$$

This motivates us to consider the one-step iterative method

$$v_{k+1} = v_k - \alpha D^{-1}(Av_k - b). \tag{9.14}$$

Here, $\alpha > 0$ is the learning rate. If $\alpha = 1$ and $D_{jj} = A_{jj}$ for $j = 1, \cdots, m$, then the above iteration is called the *Jacobi method*. If $D = I$ and A is a positive definite symmetric matrix, then the above iteration is the gradient descent method for the optimization problem of minimizing the quadratic function

$$Q(v) := \frac{1}{2} v^T A v - v^T b.$$

Let $e_k = v_k - v$ be the error vector. The iteration (9.14) is rewritten as

$$e_{k+1} = (I - \alpha D^{-1} A) e_k = (I - \alpha D^{-1} A)^k e_1. \tag{9.15}$$

Let $0 < \lambda_1 \leq \cdots \leq \lambda_m$ be the eigenvalues of $D^{-1}A$ (counting multiplicity). Denote

$$J_1(\alpha) := \rho(I - \alpha D^{-1} A) = \max_{j=1,\cdots,m} |1 - \alpha \lambda_j|. \tag{9.16}$$

If $\alpha > 0$ is sufficiently small, say $\alpha < 1/\lambda_m$, then $J_1(\alpha) < 1$ and $e_k \to 0$ as $k \to \infty$. To obtain a fast convergence rate, we shall investigate the following minimax optimization problem.

Minimax Optimization Problem I
Let $0 < \lambda_1 \leq \cdots \leq \lambda_m$ be the eigenvalues of $D^{-1}A$ (counting multiplicity). Find $\alpha > 0$ such that the objective function $J_1(\alpha)$ defined in (9.16) is minimized.

Theorem 9.2 *Let $\alpha^* = 2/(\lambda_m + \lambda_1)$. We have*

$$J_1(\alpha) > J_1(\alpha^*) = \frac{\lambda_m - \lambda_1}{\lambda_m + \lambda_1} \tag{9.17}$$

for all $\alpha > 0$ with $\alpha \neq \alpha^*$.

Proof Given $\alpha > 0$, then function $f_\alpha(\lambda) = |1 - \alpha\lambda|$ is a convex function, and its maximum value on $[\lambda_1, \lambda_m]$ is achieved at either λ_1 or λ_m. Hence, we obtain from (9.16) that

$$J_1(\alpha) = \max\{f_\alpha(\lambda_1), f_\alpha(\lambda_m)\}.$$

If $\alpha = \alpha^* = 2/(\lambda_m + \lambda_1)$, then $f_\alpha(\lambda_1) = f_\alpha(\lambda_m) = (\lambda_m - \lambda_1)/(\lambda_m + \lambda_1)$. This implies that

$$J_1(\alpha^*) = (\lambda_m - \lambda_1)/(\lambda_m + \lambda_1).$$

If $\alpha > \alpha^*$, then

$$J_1(\alpha) \geq f_\alpha(\lambda_m) = \alpha\lambda_m - 1 > \alpha^*\lambda_m - 1 = J_1(\alpha^*).$$

If $\alpha \in (0, \alpha^*)$, then

$$J_1(\alpha) \geq f_\alpha(\lambda_1) = 1 - \alpha\lambda_1 - 1 > 1 - \alpha^*\lambda_1 = J_1(\alpha^*).$$

This completes the proof. □

9.3 Two-Step Method

We generalize the one-step method (9.14) as

$$v_{k+1} = v_k - \alpha D^{-1}(Av_k - b) + \beta(v_k - v_{k-1}), \tag{9.18}$$

where $\alpha > 0$ is the learning rate, and $\beta \geq 0$ is another hyperparameter related to the moment. When $\beta = 0$, the above two-step iterative method reduces to the one-step method in (9.14). In the literature, this two-step method is also called the *heavy ball method*. Now, the error vector $e_k = v_k - v$ satisfies a second-order difference equation:

$$e_{k+1} = [(1+\beta)I - \alpha D^{-1}A]e_k - \beta e_{k-1}, \tag{9.19}$$

which can be rewritten as

$$\begin{pmatrix} e_{k+1} \\ e_k \end{pmatrix} = \begin{pmatrix} (1+\beta)I - \alpha D^{-1}A & -\beta I \\ I & 0 \end{pmatrix} \begin{pmatrix} e_k \\ e_{k-1} \end{pmatrix}. \tag{9.20}$$

Let $0 < \lambda_1 \leq \cdots \leq \lambda_m$ be the eigenvalues of $D^{-1}A$; then the eigenvalues of the block matrix

$$\begin{pmatrix} (1+\beta)I - \alpha D^{-1}A & -\beta I \\ I & 0 \end{pmatrix}$$

are

$$\frac{1 + \beta - \alpha\lambda_j \pm \sqrt{(1 + \beta - \alpha\lambda_j)^2 - 4\beta}}{2}, \quad j = 1, \cdots, m.$$

This motivates us to consider the following minimax optimization problem.

Minimax Optimization Problem II
Let $0 < \lambda_1 \leq \cdots \leq \lambda_m$ be the eigenvalues of $D^{-1}A$ (counting multiplicity). Find $\alpha > 0$ and $\beta \geq 0$ such that the objective function

$$J_2(\alpha, \beta) := \max_{j=1,\cdots,m} \rho_j(\alpha, \beta) \tag{9.21}$$

is minimized, where

$$\rho_j(\alpha, \beta) := \left| \frac{|1+\beta-\alpha\lambda_j| + \sqrt{(1+\beta-\alpha\lambda_j)^2 - 4\beta}}{2} \right| \tag{9.22}$$

is the spectral radius of the matrix

$$\begin{pmatrix} (1+\beta)I - \alpha\lambda_j & -\beta \\ 1 & 0 \end{pmatrix}.$$

Theorem 9.3 *Let*

$$\alpha^* = \left[\frac{2}{\sqrt{\lambda_m} + \sqrt{\lambda_1}} \right]^2, \quad \beta^* = \left[\frac{\sqrt{\lambda_m} - \sqrt{\lambda_1}}{\sqrt{\lambda_m} + \sqrt{\lambda_1}} \right]^2. \tag{9.23}$$

We have

$$J_2(\alpha, \beta) > J_2(\alpha^*, \beta^*) = \frac{\sqrt{\lambda_m} - \sqrt{\lambda_1}}{\sqrt{\lambda_m} + \sqrt{\lambda_1}} \tag{9.24}$$

for all $\alpha > 0$ *and* $\beta \geq 0$ *with* $(\alpha, \beta) \neq (\alpha^*, \beta^*)$.

Proof It is easy to verify that

$$\frac{1+\beta^* - \alpha^*\lambda_1}{2} = -\frac{1+\beta^* - \alpha^*\lambda_m}{2} = \sqrt{\beta^*} = \frac{\sqrt{\lambda_m} - \sqrt{\lambda_1}}{\sqrt{\lambda_m} + \sqrt{\lambda_1}},$$

and

$$J_2(\alpha^*, \beta^*) = \rho_1(\alpha^*, \beta^*) = \rho_m(\alpha^*, \beta^*) = \frac{\sqrt{\lambda_m} - \sqrt{\lambda_1}}{\sqrt{\lambda_m} + \sqrt{\lambda_1}}.$$

We consider the following four cases, respectively:

(i) If $0 < \alpha < \alpha^*$ and $\beta \geq \beta^*$, then we have

$$J_2(\alpha,\beta) \geq \rho_1(\alpha,\beta) \geq \frac{1+\beta-\alpha\lambda_1}{2} > \frac{1+\beta^*-\alpha^*\lambda_1}{2} = J_2(\alpha^*,\beta^*).$$

(ii) If $\alpha > \alpha^*$ and $0 \leq \beta \leq \beta^*$, then we have

$$J_2(\alpha,\beta) \geq \rho_m(\alpha,\beta) \geq -\frac{1+\beta-\alpha\lambda_m}{2} > -\frac{1+\beta^*-\alpha^*\lambda_m}{2} = J_2(\alpha^*,\beta^*).$$

(iii) If $0 < \alpha \leq \alpha^*$ and $0 \leq \beta < \beta^*$, then we have $(1+\beta-\alpha\lambda_1)^2 - 4\beta > 0$ and

$$\begin{aligned} J_2(\alpha,\beta) \geq \rho_1(\alpha,\beta) &= \frac{1+\beta-\alpha\lambda_1+\sqrt{(1+\beta-\alpha\lambda_1)^2-4\beta}}{2} \\ &\geq \frac{1+\beta-\alpha^*\lambda_1+\sqrt{(1+\beta-\alpha^*\lambda_1)^2-4\beta}}{2} \\ &> \frac{1+\beta^*-\alpha^*\lambda_1+\sqrt{(1+\beta^*-\alpha^*\lambda_1)^2-4\beta^*}}{2} = J_2(\alpha^*,\beta^*). \end{aligned}$$

(iv) If $\alpha \geq \alpha^*$ and $\beta > \beta^*$, then we have

$$J_2(\alpha,\beta) \geq \rho_m(\alpha,\beta) \geq \sqrt{\beta} > \sqrt{\beta^*} = J_2(\alpha^*,\beta^*).$$

A combination of the above four cases yields the desired results. □

Problems

9.1 Let r_k, e_k, and u_k be defined recursively from (9.3), (9.4), and (9.5). Prove that $r_{k+1}^T r_k = 0$, $r_{k+1}^T u_k = 0$, and $u_k^T r_k = r_k^T r_k$.

9.2 Verify (9.11) and (9.13).

9.3 Prove that if the iteration (9.14) converges, then the limit is a solution to the linear system $Av = b$.

9.4 Verify (9.15).

9.5 Given $\alpha > 0$ and $0 < \lambda_1 < \lambda_m$, prove that the function $f_\alpha(\lambda) = |1-\alpha\lambda|$ is a convex function, and its maximum value on $[\lambda_1, \lambda_m]$ is achieved at either λ_1 or λ_m.

9.6 Given $x \in \mathbb{R}^{n\times(d+1)}$ and $y \in \mathbb{R}^n$ with $\det(x^T x) \neq 0$, show that the minimizer of the error $SSE = \|xp - y\|_2^2$ can be approximated by the iteration (9.14) where $D = I$, $A = x^T x$, $b = x^T y$, and the learning rate $\alpha > 0$ is chosen to be sufficiently small.

9.7 Let $0 < \lambda_1 \leq \cdots \leq \lambda_m$ be the eigenvalues of $D^{-1}A$; prove that the eigenvalues of the block matrix

$$\begin{pmatrix} (1+\beta)I - \alpha D^{-1}A & -\beta I \\ I & 0 \end{pmatrix}$$

are

$$\frac{1+\beta-\alpha\lambda_j \pm \sqrt{(1+\beta-\alpha\lambda_j)^2-4\beta}}{2}, \quad k=1,\cdots,m.$$

9.8 Given $\alpha > 0$, $\beta \geq 0$, and $\lambda_j > 0$, show that $\rho_j(\alpha, \beta)$ defined in (9.22) is the spectral radius of the matrix

$$\begin{pmatrix} (1+\beta)I - \alpha\lambda_j & -\beta \\ 1 & 0 \end{pmatrix}.$$

9.9 Given $0 < \lambda_1 < \lambda_m$, let α^* and β^* be defined as in (9.23). For any $\lambda_j \in (\lambda_1, \lambda_m)$, prove that

$$\rho_j(\alpha^*, \beta^*) < \rho_1(\alpha^*, \beta^*) = \rho_m(\alpha^*, \beta^*) = \frac{\sqrt{\lambda_m}-\sqrt{\lambda_1}}{\sqrt{\lambda_m}+\sqrt{\lambda_1}},$$

where $\rho_j(\alpha, \beta)$ is defined as in (9.22).

9.10 Given $0 < \lambda_1 < \lambda_m$, let α^* and β^* be defined as in (9.23). Prove that the function

$$f(\beta) := 1+\beta-\alpha^*\lambda_1 + \sqrt{(1+\beta-\alpha^*\lambda_1)^2-4\beta}$$

is strictly decreasing (i.e., $f'(\beta) < 0$) for $\beta \in [0, \beta^*]$.

9.11 For any $\beta \geq 0$ and $\gamma \geq 0$, prove that $|\gamma + \sqrt{\gamma^2 - 4\beta}| \geq 2\sqrt{\beta}$.

Chapter 10
Dimensionality Reduction

Abstract This chapter presents the theory of singular value decomposition (SVD) and principal component analysis (PCA), which are fundamental tools for analyzing and approximating high-dimensional data. We begin by reviewing the Schur decomposition for square matrices and unitary transformations, laying the groundwork for understanding SVD. The chapter then introduces singular value decomposition, proving its existence, uniqueness, and the relationship between singular values and the eigenvalues of $A^T A$ and AA^T. Building on SVD, we develop principal component analysis as a method for dimensionality reduction. We define the Rayleigh quotient and formulate a sequence of optimization problems whose solutions yield the principal components of a given matrix. The chapter establishes the orthogonality of left and right singular vectors and demonstrates that the sum of the first k principal components provides the best rank-k approximation of the original matrix in the Frobenius norm.

10.1 Schur Decomposition

Recall that a matrix $U \in \mathbb{C}^{m\times m}$ is *unitary* if $UU^H = U^H U = I$, where $U^H \in \mathbb{C}^{m\times m}$ is the conjugate transpose of U defined as $U^H_{jk} = \overline{U_{kj}}$.

Theorem 10.1 *For any $A \in \mathbb{C}^{m\times m}$, there exist a unitary matrix $U \in \mathbb{C}^{m\times m}$ (i.e., $UU^H = U^H U = I$) and an upper triangular matrix $T \in \mathbb{C}^{m\times m}$ (i.e., $T_{jk} = 0$ for $1 \le k < j \le m$) such that $A = UTU^H$. We may order the diagonal terms of T such that $|T_{11}| \ge |T_{22}| \ge \cdots \ge |T_{mm}|$.*

Proof We prove by induction. The result is trivial when $m = 1$. For a general $m > 1$, assume the result is true for $m - 1$. Let λ_1 be the eigenvalue of A with the largest modulus. Let $u_1 \in \mathbb{C}^m$ be the associated eigenvector with $\|u_1\|_2 = 1$. By a standard Gram-Schmidt process, we can find $m - 1$ orthonormal vectors $u_2, \cdots, u_m \in \mathbb{C}^m$ such that $U_1 = (u_1, \cdots, u_m) \in \mathbb{C}^{m\times m}$ is unitary. Since $Au_1 = \lambda_1 u_1$, we have

X.-S. Wang, C. Wang, *Machine Learning in Data Processing*, Forum for Interdisciplinary Mathematics, https://doi.org/10.1007/978-3-032-20855-2_10

$$AU_1 = U_1 \begin{pmatrix} \lambda_1 & \alpha_1^H \\ 0 & A_2 \end{pmatrix},$$

where $\alpha_1 \in \mathbb{C}^{m-1}$ and $A_2 \in \mathbb{C}^{(m-1)\times(m-1)}$. By inductive assumption, there exist a unitary matrix $U_2 \in \mathbb{C}^{(m-1)\times(m-1)}$ and an upper triangular matrix $T_2 \in \mathbb{C}^{(m-1)\times(m-1)}$ such that $A_2 = U_2 T_2 U_2^H$. Substituting this into the above equation gives

$$A = U_1 \begin{pmatrix} \lambda_1 & \alpha_1^H \\ 0 & U_2 T_2 U_2^H \end{pmatrix} U_1^H.$$

In view of the matrix factorization

$$\begin{pmatrix} \lambda_1 & \alpha_1^H \\ 0 & U_2 T_2 U_2^H \end{pmatrix} = \begin{pmatrix} 1 & 0 \\ 0 & U_2 \end{pmatrix} \begin{pmatrix} \lambda_1 & \alpha_1^H U_2 \\ 0 & T_2 \end{pmatrix} \begin{pmatrix} 1 & 0 \\ 0 & U_2^H \end{pmatrix},$$

we obtain $A = UTU^H$, where

$$U = \begin{pmatrix} 1 & 0 \\ 0 & U_2 \end{pmatrix}$$

is unitary and

$$T = \begin{pmatrix} \lambda_1 & \alpha_1^H U_2 \\ 0 & T_2 \end{pmatrix}$$

is upper triangular. This completes the proof. □

10.2 Singular Value Decomposition

Theorem 10.2 *For any matrix $A \in \mathbb{C}^{m\times n}$, there exist a unitary matrix $U \in \mathbb{C}^{m\times m}$, a unitary matrix $V \in \mathbb{C}^{n\times n}$, and a rectangular diagonal matrix $\Sigma \in \mathbb{C}^{m\times n}$ with nonnegative diagonal terms, such that $A = U\Sigma V^H$. We may order the diagonal terms of Σ such that $\Sigma_{jj} \geq \Sigma_{kk}$ for $1 \leq j < k \leq \min\{m, n\}$.*

Proof Without loss of generality, we may assume that $m \leq n$ and A is of rank m; namely, $AA^H \in \mathbb{C}^{m\times m}$ is a positive definite Hermitian matrix. There exist a unitary matrix $U \in \mathbb{C}^{m\times m}$ and a diagonal matrix $D \in \mathbb{C}^{m\times m}$ with positive diagonal terms such that $AA^H = UD^2U^H$. In particular, we may order the diagonal terms of D as $D_{11} \geq D_{22} \geq \cdots \geq D_{mm} > 0$. Set $V_1 = A^H U D^{-1} \in \mathbb{C}^{n\times m}$. It is easily seen that $A = UDV_1^H$ and

$$V_1^H V_1 = D^{-1} U^H A A^H U^H D^{-1} = D^{-1} D^2 D^{-1} = I.$$

Finally, we set $\Sigma = (D, 0) \in \mathbb{R}^{m\times n}$ and use a Gram-Schmidt process to choose $V_2 \in \mathbb{C}^{n\times(n-m)}$ such that $V^H = (V_1^H, V_2^H) \in \mathbb{C}^{n\times n}$ is unitary. This completes the proof. □

The decomposition $A = U\Sigma V^H$ is called a *singular value decomposition*, and the nonnegative diagonal terms of Σ are called the *singular values* of A.

Remark 10.1 If $A = U\Sigma V^H$, then $AA^H = U\Sigma\Sigma^H U^H$ and $A^H A = V\Sigma^H \Sigma V^H$. Therefore, the square of any singular value of A is an eigenvalue of AA^H (or $A^H A$). In particular, if A is Hermitian, then the absolute values of the eigenvalues of A are the singular values of A. However, for a general square matrix A, the absolute value of eigenvalues and the singular values may be different.

The following proposition shows that the rectangular diagonal matrix of the singular value decomposition is unique up to a permutation of the singular values.

Proposition 10.1 *Assume* $A = U\Sigma V^H = \tilde{U}\tilde{\Sigma}\tilde{V}^H$ *are two singular value decompositions of* $A \in \mathbb{C}^{m\times n}$, *where* U *and* $\tilde{U}$ *are unitary matrices in* $\mathbb{C}^{m\times m}$, V *and* $\tilde{V}$ *are unitary matrices in* $\mathbb{C}^{n\times n}$, *and* Σ *and* $\tilde{\Sigma}$ *are rectangular diagonal matrices in* $\mathbb{C}^{m\times n}$. *If the diagonal terms of* Σ *and* $\tilde{\Sigma}$ *are ordered as* $\Sigma_{jj} \geq \Sigma_{kk}$ *and* $\tilde{\Sigma}_{jj} \geq \tilde{\Sigma}_{kk}$ *for* $1 \leq j < k \leq \min\{m, n\}$, *then we have* $\Sigma = \tilde{\Sigma}$.

Proof Without loss of generality, we may assume $m \leq n$ and rewrite $\Sigma = (D, 0)$ and $\tilde{\Sigma} = (\tilde{D}, 0)$, where D and $\tilde{D}$ are diagonal matrices in $\mathbb{C}^{m\times m}$. Since $AA^H = UD^2U^H = \tilde{U}\tilde{D}^2\tilde{U}^H$, the diagonal terms of D^2 and $\tilde{D}^2$ are the same as the eigenvalues of AA^H. By assumption, these diagonal terms are ordered as $D_{11} \geq \cdots \geq D_{mm}$ and $\tilde{D}_{11} \geq \cdots \geq \tilde{D}_{mm}$. Therefore, we have $D = \tilde{D}$ and $\Sigma = \tilde{\Sigma}$. This completes the proof. □

10.3 Principal Component Analysis

Throughout this section, we assume that a matrix $A \in \mathbb{R}^{m\times n}$ is given. By Schur decomposition, we obtain $A^T A = V\Lambda V^T$, where $V = (v_1, \cdots, v_n)$ is an orthogonal matrix (i.e., $VV^T = V^T V = I$) and Λ is a diagonal matrix with nonnegative diagonal terms $\lambda_1 \geq \lambda_2 \geq \cdots \geq \lambda_n \geq 0$.

The first Rayleigh quotient is defined as

$$R_1(v) := \frac{v^T A^T A v}{v^T v} = \frac{\|Av\|_2^2}{\|v\|_2^2}, \tag{10.1}$$

for $v \in \mathbb{R}^n$ with $v \neq 0$.

The First Optimization Problem
Find $v \in \mathbb{R}^n$ with $\|v\|_2 = 1$ such that $R_1(v) = \|Av\|_2^2$ is maximized.

Proposition 10.2 *For any normalized vector $v \in \mathbb{R}^n$, we have $R_1(v) \leq R_1(v_1) = \lambda_1$.*

Proof For any $v = c_1 v_1 + \cdots + c_n v_n \in \mathbb{R}^n$ with $c_1^2 + \cdots + c_n^2 = 1$, we have

$$R_1(v) = v^T A^T A v = \sum_{j=1}^{n} \lambda_j c_j^2 \leq \sum_{j=1}^{n} \lambda_1 c_j^2 = \lambda_1 = R_1(v_1).$$

This completes the proof. □

For each $k = 1, \cdots, \min\{m, n\}$, we denote $\sigma_k = \sqrt{\lambda_k}$ to be the k-th singular value. The right eigenvector v_k of $A^T A$ associated with the eigenvalue λ_k is called the *right singular vector* of A associated with σ_k. If $\sigma_k > 0$, we also define $u_k = (Av_k)/\sigma_k \in \mathbb{R}^m$ as the *left singular vector* of A associated with σ_k.

Lemma 10.1 *For each $k = 1, \cdots, \min\{m, n\}$ with $\sigma_k > 0$, the left singular vectors $u_1, \cdots, u_k$ are orthonormal vectors in $\mathbb{R}^m$.*

Proof For any $j = 1, \cdots, k$, by definition, we have $u_j^T u_j = (v_j^T A^T A v_j)/\lambda_j = 1$, and hence u_j is a normalized vector. If $1 \leq j < k \leq \min\{m, n\}$ and $\sigma_k > 0$, then $u_j^T u_k = (v_j^T A^T A v_k)/(\sigma_j \sigma_k) = 0$. Therefore, by induction, for each $k = 1, \cdots, \min\{m, n\}$ with $\sigma_k > 0$, the vectors $u_1, \cdots, u_k$ are orthonormal vectors. □

For each $k = 1, \cdots, \min\{m, n\}$, we define the k-th principal component of A as

$$P_k := A v_k v_k^T. \tag{10.2}$$

We can regard P_k as the projection of A on v_k. The k-th Rayleigh quotient is given by

$$R_k(v) := \frac{v^T A_k^T A_k v}{v^T v} = \frac{\|A_k v\|_2^2}{\|v\|_2^2}, \tag{10.3}$$

for any nonzero vector $v \in \mathbb{R}^n$, where

$$A_k := A - \sum_{j=1}^{k-1} P_j = A - \sum_{j=1}^{k-1} A v_j v_j^T. \tag{10.4}$$

If $v \in \mathbb{R}^n$ satisfies $\|v\|_2 = 1$ and $v^T v_j = 0$ for $j = 1, \cdots, k-1$, then $A_k v = Av$ and $R_k(v) = \|A_k v\|_2^2 = \|Av\|_2^2$.

The k-th Optimization Problem
For each $k = 2, \cdots, \min\{m, n\}$, find $v \in \mathbb{R}^n$ with $\|v\|_2 = 1$ and $v^T v_j = 0$ for $j = 1, \cdots, k-1$, such that $R_k(v) = \|A_k v\|_2^2 = \|Av\|_2^2$ defined as in (10.3) is maximized.

Proposition 10.3 *For each $k = 2, \cdots, \min\{m, n\}$, we have $R_k(v) \leq R_k(v_k) = \lambda_k$ for any normalized vector $v \in \mathbb{R}^n$ with $v^T v_j = 0$ for $j = 1, \cdots, k-1$.*

Proof For any $v = c_1 v_1 + \cdots + c_n v_n \in \mathbb{R}^n$ with $c_1^2 + \cdots + c_n^2 = 1$ and $v^T v_j = 0$ for $j = 1, \cdots, k-1$, we have $c_1 = \cdots = c_{k-1} = 0$, and

$$R_k(v) = v^T A^T A v = \sum_{j=k}^{n} \lambda_j c_j^2 \leq \sum_{j=k}^{n} \lambda_k c_j^2 = \lambda_k = R_k(v_k).$$

This completes the proof. □

Theorem 10.3 *If $\sigma_k = 0$ for some $k = 1, \cdots, \min\{m, n\}$, then $A_k = 0$; namely,*

$$A = \sum_{j=1}^{k-1} P_j. \tag{10.5}$$

If $\sigma_k > 0$ for all $k = 1, \cdots, \min\{m, n\}$, then we have

$$A = \sum_{j=1}^{\min\{m,n\}} P_j. \tag{10.6}$$

Proof From (10.4), we have $A_k v_j = 0$ for any $1 \leq j < k$. If $\sigma_k = 0$ for some $k = 1, \cdots, \min\{m, n\}$, then Proposition 10.3 implies that $A_k v_j = 0$ for all $k \leq j \leq n$. Therefore, $A_k v = 0$ for any $v = c_1 v_2 + \cdots + c_n v_n \in \mathbb{R}^n$. Consequently, $A_k = 0$, and (10.5) is proved.

Now, we assume that $\sigma_k > 0$ for all $k = 1, \cdots, \min\{m, n\}$. If $m \geq n$, then by (10.4), we obtain $A_{n+1} v_j = 0$ for all $j = 1, \cdots, n$, which implies $A_{n+1} = 0$. If $m \leq n$, then from $u_j = (Au_j)/\sigma_j$ we obtain $\sigma_j u_j^T A_{m+1} = v_j^T A^T A_{m+1} = 0$ for all $j = 1, \cdots, m$, which implies that $A_{m+1} = 0$. This proves (10.6). □

The following lemma is a generalization of triangle inequality and Pythagorean theorem.

Lemma 10.2 *Let $w_1, \cdots, w_k \in \mathbb{R}^n$ be an orthonormal basis of a subspace $W \subset \mathbb{R}^n$. For any $u \in \mathbb{R}^n$ and $w \in W$, we have*

$$\|u - w\|_2^2 \geq \|u\|_2^2 - \sum_{j=1}^{k}(u^T w_j)^2,$$

where the equality holds if and only if

$$w = \sum_{j=1}^{k}(u^T w_j)w_j$$

is the projection of u *on* W.

Proof If $k = 1$, then

$$\|u - cw_1\|_2^2 = c^2 - 2(u^T w_1)c + u^T u \geq \|u\|_2^2 - (u^T w_1)^2,$$

where the equality holds if and only if $c = u^T w_1$. If $k \geq 2$, by induction, we obtain

$$\|u - \sum_{j=1}^{k} c_j w_j\|_2^2 \geq \|u - \sum_{j=1}^{k-1} c_j w_j\|_2^2 - (u^T w_k)^2 \geq \|u\|_2^2 - \sum_{j=1}^{k}(u^T w_j)^2,$$

where the equality holds if and only if $c_j = u^T w_j$ for all $j = 1, \cdots, k$. This completes the proof. □

Lemma 10.3 *Let* W *be a subspace of* $\mathbb{R}^n$ *with dimension* $k \leq n$. *For any vectors* $e_1, \cdots, e_{k-1} \in \mathbb{R}^n$, *there exists* $w \in W$ *such that* $\|w\|_2 = 1$ *and* $w^T e_j = 0$ *for all* $j = 1, \cdots, k-1$.

Proof Let $w_1, \cdots, w_k$ be an orthonormal basis of W. We define a matrix

$$B := (e_1, \cdots, e_{k-1})^T (w_1, \cdots, w_k) \in \mathbb{R}^{(k-1)\times k},$$

namely, $B_{jl} = e_j^T w_l$ for $j = 1, \cdots, k-1$ and $l = 1, \cdots, k$. The column vectors of

$$B^T = (w_1, \cdots, w_k)^T (e_1, \cdots, e_{k-1}) \in \mathbb{R}^{(k-1)\times k}$$

are denoted by $b_1, \cdots, b_{k-1} \in \mathbb{R}^k$; namely, $B^T = (b_1, \cdots, b_k)$ with

$$b_j = (w_1, \cdots, w_k)^T e_j, \quad j = 1, \cdots, k-1.$$

Since the subspace spanned by these vectors has a dimension less than k, there exists a vector $c = (c_1, \cdots, c_k)^T \in \mathbb{R}^k$ such that $\|c\|_2 = 1$ and $c^T b_j = 0$ for all $j = 1, \cdots, k-1$. This implies that $Bc = 0$. By choosing

$$w = w_1 c_1 + \cdots + w_k c_k \in \mathbb{R}^n,$$

we obtain

$$e_j^T w = B_{j1}c_1 + \cdots + B_{jk}c_k = (Bc)_j = 0$$

for each $j = 1, \cdots, k-1$. Moreover, $\|w\|_2^2 = c_1^2 + \cdots + c_k^2 = 1$. The proof is completed. □

The following theorem demonstrates that the sum of first k principal components provides the best approximation to A among all matrices of rank k with respect to the Frobenius norm. Recall that the Frobenius norm of $B \in \mathbb{R}^{m\times n}$ is the same as the l^2-norm of B when viewed as a vector in $\mathbb{R}^{mn}$; namely,

$$\|B\|_F := \sqrt{\sum_{j=1}^{m}\sum_{k=1}^{n} |B_{jk}|^2}. \tag{10.7}$$

Theorem 10.4 *For any $B \in \mathbb{R}^{m\times n}$ with* rank$(B) = k$, *we have*

$$\|A - B\|_F^2 \geq \sum_{j=k+1}^{n} \lambda_j,$$

where the equality holds if $B = P_1 + \cdots + P_k$ is the sum of first k principal components.

Proof Denote the column vectors of A^T and B^T by $a_1, \cdots, a_m \in \mathbb{R}^n$ and $b_1, \cdots, b_m \in \mathbb{R}^n$, respectively. Then we have

$$\|A - B\|_F^2 = \sum_{l=1}^{m} \|a_l - b_l\|_2^2.$$

Since rank$(B) = k$, the subspace $W = \text{span}\{b_1, \cdots, b_m\}$ has a dimension k. Recall that $v_1, \cdots, v_n$ are eigenvectors of $A^T A$ corresponding to $\lambda_1, \cdots, \lambda_n$, respectively. These vectors form an orthonormal basis in $\mathbb{R}^n$. By repeated applications of Lemma 10.3, we can find an orthonormal basis $w_1, \cdots, w_k$ in W such that $w_j^T v_l = 0$ for any $1 \leq l < j \leq k$. For each $l = 1, \cdots, m$, it follows from Lemma 10.2 that

$$\|a_l - b_l\|_2^2 \geq \|a_l\|_2^2 - \sum_{j=1}^{k} (a_l^T w_j)^2,$$

where the equality holds if and only if

$$b_l = \sum_{j=1}^{k} (a_l^T w_j) w_j$$

is the projection of a_l on W. Hence, we obtain

$$\|A - B\|_F^2 \geq \sum_{l=1}^{m} \|a_l\|_2^2 - \sum_{l=1}^{m}\sum_{j=1}^{k}(a_l^T w_j)^2 = \|A\|_F^2 - \sum_{j=1}^{k} \|Aw_j\|_2^2.$$

For each $j = 1, \cdots, k$, on account of Propositions 10.2 and 10.3, we have $\|Aw_j\|_2^2 \leq \|Av_j\|_2^2 = \lambda_j$. Consequently, we obtain

$$\|A - B\|_F^2 \geq \sum_{j=1}^{n} \lambda_j - \sum_{j=1}^{k} \lambda_j = \sum_{j=k+1}^{n} \lambda_j,$$

where the equality holds if

$$b_l = \sum_{j=1}^{k}(a_l^T v_j)v_j;$$

namely,

$$B = \sum_{j=1}^{k} Av_j v_j^T = \sum_{j=1}^{k} P_j.$$

This completes the proof. □

Problems

10.1 For any $u_1 \in \mathbb{C}^n$ with $\|u_1\|_2 = 1$, show that there exist normalized orthonormal vectors $u_2, \cdots, u_m \in \mathbb{C}^m$ such that $U_1 = (u_1, \cdots, u_m) \in \mathbb{C}^{m\times m}$ is unitary.

10.2 Verify

$$\begin{pmatrix} \lambda_1 & \alpha_1^H \\ 0 & U_2 T_2 U_2^H \end{pmatrix} = \begin{pmatrix} 1 & 0 \\ 0 & U_2 \end{pmatrix} \begin{pmatrix} \lambda_1 & \alpha_1^H U_2 \\ 0 & T_2 \end{pmatrix} \begin{pmatrix} 1 & 0 \\ 0 & U_2^H \end{pmatrix}.$$

10.3 If $A \in \mathbb{C}^{m\times m}$ is normal, i.e., $AA^H = A^H A$, prove that A is unitarily diagonalizable; namely, $A = UDU^H$ for some unitary matrix U and diagonal matrix D.

10.4 If $A \in \mathbb{R}^{m\times m}$ is real symmetric (i.e., $A^T = A$), prove that there exist an orthogonal matrix $V \in \mathbb{R}^{m\times m}$ (i.e., $VV^T = V^T V = I$) and a diagonal matrix $D \in \mathbb{R}^{m\times m}$ such that $A = VDV^T$.

10.5 Assume $m \leq n$. For any $V_1 \in \mathbb{C}^{n\times m}$ with $V_1^H V_1 = I$, prove that there exists $V_2 \in \mathbb{C}^{n\times(n-m)}$ such that $V^H = (V_1^H, V_2^H) \in \mathbb{C}^{n\times n}$ is unitary.

10.6 Compute the eigenvalues and singular values of the matrix

$$A = \begin{pmatrix} 1 & 2 \\ 0 & 1 \end{pmatrix}.$$

10.7 Let $A \in \mathbb{R}^{m\times n}$ with the first Rayleigh quotient $R_1(v)$ defined as in (10.1). For any nonzero vector $v \in \mathbb{R}^n$ and its normalization $w = v/\|v\|_2$. Prove that $R_1(v) = R_1(w) = \|Aw\|_2^2$.

10.8 Given $A \in \mathbb{R}^{m\times n}$, let $A_k \in \mathbb{R}^{m\times n}$ defined as in (10.4). Prove that

$$A_k^T A_k = A^T A - \sum_{j=1}^{k-1} \lambda_j v_j v_j^T.$$

10.9 Given any vectors $b_1, \cdots, b_{k-1} \in \mathbb{R}^k$, prove that there exists a vector $u \in \mathbb{R}^k$ such that $\|u\|_2 = 1$ and $u^T b_j = 0$ for all $j = 1, \cdots, k-1$.

10.10 Given $A \in \mathbb{R}^{m\times n}$, let $a_1, \cdots, a_m \in \mathbb{R}^n$ be the column vectors of A^T. Prove that

$$\|A^T\|_F^2 = \|A\|_F^2 = \sum_{j=1}^{m} \|a_j\|_2^2.$$

Appendix A
Related Topics

In this appendix, we review some related topics and results from calculus and linear algebra.

A.1 Calculus

We say an infinite sequence $\{x_n\}_{n=1}^{\infty} \subset \mathbb{R}$ converges to c if for any $\varepsilon > 0$, there exists $N \in \mathbb{N}$ such that $|x_n - c| < \varepsilon$ for all $n > N$. In this case, we denote

$$\lim_{n \to \infty} x_n = c.$$

The sequence $\{x_n\}_{n=1}^{\infty} \subset \mathbb{R}$ is called a Cauchy sequence if for any $\varepsilon > 0$, there exists $N \in \mathbb{N}$ such that $|x_n - x_m| < \varepsilon$ for all $n > N$ and $m > N$.

Theorem A.1 *The real line $\mathbb{R}$ is complete; namely, the following statements are equivalently valid:*

(i) Every Cauchy sequence converges.
(ii) Every bounded and monotone sequence is convergent.
(iii) Every bounded sequence has a convergent subsequence.
(iv) Every non-empty bounded set has a supremum (least upper bound) and an infimum (greatest lower bound)

Given a non-empty set $A \subset \mathbb{R}$, we say $a \in \mathbb{R}$ is a limit point of A if there exists an infinite sequence $\{x_n\}_{n=1}^{\infty} \subset A$ that converges to a: $x_n \to a$ as $n \to \infty$. The closure of A, denoted by $\bar{A}$, is the set containing all limit points of A. Since any point in A is a limit point (by choosing $x_n = a$ for all $n = 1, 2, \cdots$), we have $A \subset \bar{A}$. We say A is closed if $A = \bar{A}$.

X.-S. Wang, C. Wang, *Machine Learning in Data Processing*, Forum for Interdisciplinary Mathematics, https://doi.org/10.1007/978-3-032-20855-2

Theorem A.2 *A set $A \subset \mathbb{R}$ is a closed and bounded set if and only if it is compact; namely, any sequence in A has a subsequence that converges to a point in A.*

Let f be a (real-valued) function defined on an open interval $I = (a, b) \subset \mathbb{R}$, where a could be either finite or $-\infty$, and b could be either finite or ∞. Given $c \in I$, we say $f(x) \to L$ as $x \to c$ if for any small $\varepsilon > 0$, there exists $\delta > 0$ such that $0 < |x - c| < \delta$ implies $|f(x) - f(c)| < \varepsilon$. The value L is called the limit of $f(x)$ at c. The one-sided limits of $f(x)$ at c can be defined in a similar manner. We say $f(x) \to L$ as $x \to c^+$ (respectively, $x \to c^-$) if for any small $\varepsilon > 0$, there exists $\delta > 0$ such that $0 < x - c < \delta$ (respectively, $-\delta < x - c < 0$) implies $|f(x) - f(c)| < \varepsilon$. We use

$$\lim_{x \to c} f(x), \quad \lim_{x \to c^+} f(x), \quad \lim_{x \to c^-} f(x),$$

respectively, to denote the limit, right limit, and left limit of $f(x)$ at c. It is evident that the limit exists if and only if both one-sided limits exist and equal. We say f is continuous at c if

$$\lim_{x \to c} f(x) = f(c).$$

One can prove that f is continuous at c if and only if for any convergent sequence $\{x_n\}_{n=1}^{\infty} \subset \mathbb{R}$ with $\lim_{n \to \infty} x_n = c$, we have $\lim_{n \to \infty} f(x_n) = f(c)$. We say f is differentiable at c if the limit

$$\lim_{x \to c} \frac{f(x) - f(c)}{x - c}$$

exists, and in this case, we denote the value of the limit by $f'(c)$ and call it the derivative of f at c. If f is continuous for each $x \in I$, then we say f is continuous on I. The set of all continuous functions on I is denoted by $C(I) = C(I, \mathbb{R})$. When the end points a and b are finite, we say f is continuous on $\bar{I} = [a, b]$ if f is continuous on $I = (a, b)$ and

$$\lim_{x \to a^+} f(x) = f(a), \quad \lim_{x \to b^-} f(x) = f(b).$$

We say f is differentiable on I if f is differentiable at each $x \in \mathrm{i}$ and $f' \in C(I)$; in this case we say $f \in C^1(I) = C^1(I, \mathbb{R})$. The higher order derivatives are defined as derivatives of derivatives. We say $f \in C^n(I)$ if the derivative functions $f', f'', \cdots, f^{(n)}$ are continuous on I.

Theorem A.3 *Let $I = (a, b)$ with $a < b$, and assume $f \in C(\bar{I})$. We have the following results:*

(i) The set $f(\bar{I}) := \{f(x) : x \in \bar{I}\}$ is compact (i.e., bounded and closed).

(ii) *(Extreme Value Theorem) Let M and m be the supremum and infimum of $f(\bar{I})$, respectively. There exist $x_M, x_m \in \bar{I}$ such that $f(x_M) = M$ and $f(x_m) = m$.*
(iii) *(Intermediate Value Theorem) For any $r \in (m, M)$, where M and m are given as above, there exists $c \in (a, b)$ such that $f(c) = r$.*
(iv) *(Mean Value Theorem) If further $f \in C^1(I)$, then there exists $\xi \in (a, b)$ such that $f'(\xi) = [f(b) - f(a)]/(b - a)$.*

Differentiation is a linear operator from $C^1(I)$ to $C(I)$; namely

$$(c_1 f_1 + c_2 f_2)' = c_1 f_1' + c_2 f_2', \tag{A.1}$$

for any $c_1, c_2 \in \mathbb{R}$ and $f_1, f_2 \in C^1(I)$. The product rule and the quotient rule of differentiation are

$$(fg)' = f'g + fg', \quad (f/g)' = (f'g - fg')/g^2, \tag{A.2}$$

for $f, g \in C^1(I)$. If $g \in C^1(I)$ and $f \in C^1(g(I))$, we also have the chain rule

$$[f(g(x))]' = f'(g(x))g'(x). \tag{A.3}$$

Given $f \in C(I)$, we say $F \in C^1(I)$ is an antiderivative of f if $F'(x) = f(x)$ for all $x \in (a, b)$. Note that if $F(x)$ is an antiderivative of $f(x)$, so is $F(x) + C$ for any constant C. If further $F \in C(\bar{I})$, then we have the fundamental theorem of calculus

$$\int_a^b f(x)dx = F(x)\Big|_a^b = F(b) - F(a), \tag{A.4}$$

where the left-hand side is called a definite integral of $f(x)$ over the interval $I = (a, b)$. If $f, g \in C^1(I) \cap C(\bar{I})$, then the product rule of differentiation (A.2) implies the useful formula of integration by parts:

$$\int_a^b f'(x)g(x)dx = f(x)g(x)\Big|_a^b - \int_a^b f(x)g'(x)dx. \tag{A.5}$$

Theorem A.4 (Taylor's Theorem) *Let $f \in C^{(n+1)}(I)$ with $I = (a, b)$. For any $x, x_0 \in (a, b)$, we have $f(x) = P_n(x) + R_n(x)$, where*

$$P_n(x) = \sum_{k=0}^{n} \frac{f^{(k)}(x_0)}{k!}(x - x_0)^k \tag{A.6}$$

is the Taylor polynomial, and

$$R_n(x) = \int_{x_0}^{x} \frac{(x - t)^n}{n!} f^{(n+1)}(t)dt \tag{A.7}$$

is the remainder/error term. Moreover, there exists $\xi \in (a, b)$ *(depending on* x *and* x_0*) such that*

$$R_n(x) = \frac{(x - x_0)^{n+1}}{(n+1)!} f^{(n+1)}(\xi). \tag{A.8}$$

A.2 Linear Algebra

A matrix $A \in \mathbb{C}^{m \times n}$ is a set of $m \times n$ complex numbers with the following structure:

$$A = \begin{pmatrix} A_{11} & \cdots & A_{1n} \\ \vdots & & \vdots \\ A_{m1} & \cdots & A_{mn} \end{pmatrix}.$$

The Hermitian of a matrix $A \in \mathbb{C}^{m \times n}$ is denoted as $A^H \in \mathbb{C}^{n \times m}$ such that $A_{ij}^H = \overline{A_{ji}}$ (namely, A_{ij}^H is the complex conjugate of A_{ji}) for all $i = 1, \cdots, n$ and $j = 1, \cdots, m$. A square matrix $A \in \mathbb{C}^{n \times n}$ is Hermitian if $A = A^H$. The transpose of a real matrix $A \in \mathbb{R}^{m \times n}$ is denoted as $A^T \in \mathbb{R}^{n \times m}$ such that $A_{ij}^T = A_{ji}$ for all $i = 1, \cdots, n$ and $j = 1, \cdots, m$. A real square matrix $A \in \mathbb{R}^{n \times n}$ is symmetric if $A = A^T$. The matrix multiplication of $A \in \mathbb{C}^{m \times n}$ and $B \in \mathbb{C}^{n \times d}$ is defined as $AB \in \mathbb{C}^{m \times d}$ such that

$$(AB)_{ij} = \sum_{k=1}^{n} A_{ik} B_{kj},$$

for all $i = 1, \cdots, m$ and $j = 1, \cdots, d$. An identity matrix $I \in \mathbb{C}^{n \times n}$ is a special matrix defined as

$$I_{ij} = \delta_{ij} = \begin{cases} 1, & i = j, \\ 0, & i \neq j, \end{cases}$$

for $i, j = 1, \cdots, n$. It is easily seen that $AI = A$ and $IB = B$ for any $A \in \mathbb{C}^{m \times n}$ and $B \in \mathbb{C}^{n \times d}$. We say a square matrix $A \in \mathbb{C}^{n \times n}$ is invertible/nonsingular if there exists $B \in \mathbb{C}^{n \times n}$ such that $AB = BA = I$. A square matrix $A \in \mathbb{C}^{n \times n}$ is unitary if $AA^H = A^H A = I$. Moreover, we say a square matrix $A \in \mathbb{C}^{n \times n}$ is normal if $AA^H = A^H A$. The diagonal, Hermitian, and unitary matrices are special examples of normal matrices. The determinant of a square matrix $A \in \mathbb{C}^{n \times n}$ is defined recursively as

$$\det(A) = \begin{cases} A_{11}, & n = 1, \\ \sum_{j=1}^{n} (-1)^{j-1} A_{1j} \det(A^{(1,j)}), & n > 1, \end{cases}$$

where $A^{(i,j)} \in \mathbb{C}^{(n-1)\times(n-1)}$ is the submatrix obtained by deleting the i-th row and the j-th column in A. The trace of a square matrix $A \in \mathbb{C}^{n\times n}$ is defined as the sum of diagonal terms:

$$\mathrm{tr}(A) := \sum_{j=1}^{n} A_{jj}.$$

For any $A, B \in \mathbb{C}^{n\times n}$, it follows from the definition that

$$\mathrm{tr}(AB) = \mathrm{tr}(BA) = \sum_{j=1}^{n} \sum_{k=1}^{n} A_{jk} B_{kj}.$$

A (column) vector $v \in \mathbb{C}^n = \mathbb{C}^{n\times 1}$ is a special matrix. We say a set of vectors $v_1, \cdots, v_n \in \mathbb{C}^n$ is linearly dependent if and only if there exist constants $c_1, \cdots, c_n \in \mathbb{C}$, not identically zero, such that

$$c_1 v_1 + \cdots + c_n v_n = 0 \in \mathbb{C}^n.$$

In other words, the vectors $v_1, \cdots, v_n$ are linearly independent if and only if the above equation has only one trivial solution $c_1 = \cdots = c_n = 0$. If the vectors $v_1, \cdots, v_k \in \mathbb{C}^n$ are linearly independent, then we can define a k-dimensional subspace $V_k \in \mathbb{C}^n$ spanned by these vectors as follows:

$$V_k = \mathrm{span}\{v_1, \cdots, v_k\} := \{c_1 v_1 + \cdots + c_k v_k,\ \ c_1, \cdots, c_k \in \mathbb{C}\}.$$

In particular, if $k = n$ and $v_1, \cdots, v_n \in \mathbb{C}^n$ are linearly independent, then $V_n = \mathrm{span}\{v_1, \cdots, v_n\} = \mathbb{C}^n$. We call $v_1, \cdots, v_k$ a basis of V_k in the sense that any vector $v \in V_k$ can be expressed as a linear combination of the basis. If these vectors are orthogonal (i.e., $v_j^H v_l = 0$ for $1 \le j < l \le k$), then the basis is orthogonal. Moreover, we say the basis is orthonormal if in addition to orthogonality, the vectors are normalized as $\|v_j\|_2 = 1$ for $j = 1, \cdots, k$.

Theorem A.5 *Let $A \in \mathbb{C}^{n\times n}$. The following statements are equivalent:*

(i) *A is invertible (nonsingular).*
(ii) $\det(A) \neq 0$.
(iii) *The linear homogeneous equation $Ax = 0 \in \mathbb{C}^n$ does not have any nontrivial solution $x \neq 0$ in $\mathbb{C}^n$.*
(iv) *The linear inhomogeneous equation $Ax = b$ has a unique solution $x \in \mathbb{C}^n$ for any given $b \in \mathbb{C}^n$.*

(v) *The columns of A, considering as a set of n column vectors, are linearly independent in C^n.*

The characteristic polynomial of a square matrix $A \in \mathbb{C}^{n\times n}$ is $p_n(\lambda) = \det(\lambda I - A) = \lambda^n + \cdots$, whose zeros are called the eigenvalues of A. According to the fundamental theorem of algebra, there are n eigenvalues (counting multiplicity), denoted by $\lambda_1, \cdots, \lambda_n$. The spectral radius of A is the largest modulus of the eigenvalues:

$$\rho(A) = \max_j |\lambda_j|.$$

For each eigenvalue λ of A, there exists a nonzero eigenvector $x \in \mathbb{C}^n$ such that $Ax = \lambda x$. If $A \in \mathbb{C}^{n\times n}$ is Hermitian, then $(x^H Ax)^H = x^H Ax \in \mathbb{R}$ for any $x \in \mathbb{C}^n$. Hence, the eigenvalues of a Hermitian matrix are real. We say a Hermitian matrix $A \in \mathbb{C}^{n\times n}$ is positive (nonnegative) definite if $x^H Ax > 0$ ($x^H Ax \geq 0$) for all $x \in \mathbb{C}^n$ with $x \neq 0$. A Hermitian matrix $A \in \mathbb{C}^{n\times n}$ is positive (nonnegative) definite if and only if all eigenvalues of A are positive (nonnegative). A nonnegative definite Hermitian matrix is also called positive semidefinite in the literature. A Hermitian matrix $A \in \mathbb{C}^{n\times n}$ is nonnegative definite if and only if there exists $B \in \mathbb{C}^{n\times n}$ such that $A = B^H B$. If further A is positive definite, then B is nonsingular/invertible.

A vector norm $\|\cdot\|$ on $\mathbb{C}^n$ is a function satisfying the following three conditions:

(i) (Positivity) $\|u\| > 0$ for all $u \in \mathbb{C}^n$ with $u \neq 0$.
(ii) (Homogeneity) $\|cu\| = |c|\|u\|$ for all $c \in C$ and $u \in \mathbb{C}^n$.
(iii) (Triangle inequality) $\|u + v\| \leq \|u\| + \|v\|$ for all $u, v \in \mathbb{C}^n$.

The most commonly used vector norms are l^p-norm with $p = 1, 2, \infty$:

$$\|u\|_1 := \sum_{i=1}^{n} |u_i|, \quad \|u\|_2 := \sqrt{\sum_{i=1}^{n} |u_i|^2}, \quad \|u\|_\infty := \max_i |u_i|.$$

An inner product $(\cdot, \cdot)$ is a bilinear function on $\mathbb{C}^n \times \mathbb{C}^n$ satisfying the following three conditions:

(i) (Positivity) $(u, u) > 0$ for all $u \in \mathbb{C}^n$ with $u \neq 0$.
(ii) (Symmetry) $(u, v) = \overline{(v, u)}$ for all $u, v \in \mathbb{C}^n$.
(iii) (Linearity) $(au+bv, w) = a(u, w)+b(v, w)$ for all $a, b \in \mathbb{C}$ and $u, v, w \in \mathbb{C}^n$.

It can be verified that $\|u\| = \sqrt{(u, u)}$ is a norm on $\mathbb{C}^n$.

Two norms $\|\cdot\|_\alpha$ and $\|\cdot\|_\beta$ on $\mathbb{C}^n$ are equivalent if there exist positive constants $C_1 > 0$ and $C_2 > 0$ such that $C_1\|u\|_\alpha \leq \|u\|_\beta \leq C_2\|u\|_\alpha$ for all $u \in \mathbb{C}^n$.

Theorem A.6 *All norms on $\mathbb{C}^n$ are equivalent and continuous. In particular, we have $\|u\|_\infty \leq \|u\|_2 \leq \sqrt{n}\|u\|_\infty$, $\|u\|_\infty \leq \|u\|_1 \leq n\|u\|_\infty$, and $\|u\|_2 \leq \|u\|_1 \leq \sqrt{n}\|u\|_2$ for all $u \in \mathbb{C}^n$.*

A matrix norm $\|\cdot\|$ on $\mathbb{C}^{m\times n}$ is a function satisfying the following three conditions:

(i) (Positivity) $\|A\| > 0$ for all $A \in \mathbb{C}^{m\times n}$ with $A \neq 0$.
(ii) (Homogeneity) $\|cA\| = |c|\|A\|$ for all $c \in C$ and $A \in \mathbb{C}^{m\times n}$.
(iii) (Triangle inequality) $\|A + B\| \leq \|A\| + \|B\|$ for all $A, B \in \mathbb{C}^{m\times n}$.

A typical matrix norm is the Frobenius norm, which is defined by treating a matrix in $\mathbb{C}^{m\times n}$ as a vector of dimension mn and then applying the l^2-norm in the mn-dimensional vector space:

$$\|A\|_F := \sqrt{\sum_{j=1}^{m}\sum_{k=1}^{n}|A_{jk}|^2} = \sqrt{\mathrm{tr}(A^H A)} = \sqrt{\mathrm{tr}(AA^H)}, \quad A \in \mathbb{C}^{m\times n}.$$

In many cases, we mainly focus on matrix norms of square matrices. In this case, we always scale the norm such that it satisfies

$$\|AB\| \leq \|A\|\|B\|, \tag{A.9}$$

for all $A, B \in \mathbb{C}^{n\times n}$. Let λ be an eigenvalue of $A \in \mathbb{C}^{n\times n}$ with a nonzero eigenvector $x \in \mathbb{C}^n$. Denote $B = (x, \cdots, x) \in \mathbb{C}^{n\times n}$. We have

$$|\lambda|\|B\| = \|\lambda B\| = \|AB\| \leq \|A\|\|B\|,$$

which implies $|\lambda| \leq \|A\|$. Since λ is an arbitrary eigenvalue, we have $\rho(A) \leq \|A\|$. This implies that the spectral radius is a lower bound of the (scaled) matrix norm. On the other hand, for any $A \in \mathbb{C}^{n\times n}$ and $\varepsilon > 0$, there exists a matrix norm (depending on A and ε) such that $\|A\| < \rho(A) + \varepsilon$.

Given any vector norm $\|\cdot\|$ on $\mathbb{C}^n$, there exists a natural/induced matrix norm $\|\cdot\|$ on $C^{n\times n}$ that is defined by

$$\|A\| = \sup_{u\in\mathbb{C}^n,\ u\neq 0} \frac{\|Au\|}{\|u\|} = \sup_{v\in\mathbb{C}^n,\ \|v\|=1} \|Av\|,$$

for any $A \in C^{n\times n}$. The most commonly used matrix norms are l^p-norms with $p = 1, 2, \infty$:

$$\|A\|_1 := \sup_{v\in\mathbb{C}^n,\ \|v\|_1=1} \|Av\|_1 = \max_{1\leq k\leq n} \sum_{j=1}^{n} |A_{jk}|,$$

$$\|A\|_2 := \sup_{v\in\mathbb{C}^n,\ \|v\|_2=2} \|Av\|_2 = \sqrt{\rho(A^H A)},$$

$$\|A\|_\infty := \sup_{v\in\mathbb{C}^n,\ \|v\|_\infty=1} \|Av\|_\infty = \max_{1\le j\le n} \sum_{k=1}^{n} |A_{jk}|.$$

Theorem A.7 *Let $A \in C^{n\times n}$. The following statements are equivalent:*

(i) $\rho(A) < 1$.
(ii) $\lim_{k\to\infty} A^k = 0$.
(iii) $\lim_{k\to\infty} A^k v = 0$ *for any* $v \in \mathbb{C}^n$.
(iv) There exists a matrix norm $\|\cdot\|$ *such that* $\|A\| < 1$.
(v) The series $\sum_{k=0}^{\infty} A^k$ *converges.*

Appendix B
Hints to Selected Exercise Problems

We provide solution hints to some of the exercise problems.

1.6 The general solution to the second-order difference equation

$$D_{n+1} = 2\cos\theta D_n - D_{n-1}$$

is

$$D_n = ae^{i\theta} + be^{-i\theta},$$

where a and b are constants determined by the initial values.

1.7 If $x_j = x_k$ for some $j \neq k$, then $\det(B) = 0$. Hence, $\prod_{j\neq k}(x_j - x_k)$ is a factor of $\det(B)$. On the other hand, $\det(B)$ is a polynomial in $(x_1, \cdots, x_n) \in \mathbb{R}^n$ of degree $n(n-1)/2$. Therefore,

$$\det(B) = c \prod_{1\le k<j\le n} (x_j - x_k),$$

where c is a constant.

2.4 Define $f(t) := J(p+t(q-p))$ for $t \in [0, 1]$, and then apply Taylor's theorem.

3.7 Fix $u \in I$. For any sufficiently small $\varepsilon > 0$, the function $g(h) := [f(u+h) - f(u)]/h$ is monotone and bounded for $h \in (0, \varepsilon]$ and also for $h \in [-\varepsilon, 0)$.

4.7 Apply Taylor's theorem to the function $f(\alpha) := J(p - \alpha J'(p))$.

5.3 Note that $(x_1 \vee x_2) \wedge x_3 = 1$ if and only if $x_1 \vee x_2 = 1$ and $x_3 = 1$, which is equivalent to $x_1 \wedge x_3 = 1$ or $x_2 \wedge x_3 = 1$. The other identity can be proved in a similar manner.

6.1 If $i \neq k$, then C_{ij} is independent of A_{kl}. If $i = k$, then from

X.-S. Wang, C. Wang, *Machine Learning in Data Processing*, Forum for Interdisciplinary Mathematics, https://doi.org/10.1007/978-3-032-20855-2

$$C_{ij} = \sum_{l=1}^{m} A_{il} B_{lj}$$

we obtain

$$\frac{\partial C_{ij}}{A_{il}} = B_{lj}.$$

The other identity follows similarly.

9.11 If $\gamma \geq 2\sqrt{\beta}$, then $|\gamma + \sqrt{\gamma^2 - 4\beta}| = \gamma + \sqrt{\gamma^2 - 4\beta} \geq \gamma \geq 2\sqrt{\beta}$. If $\gamma < 2\sqrt{\beta}$, then $|\gamma + \sqrt{\gamma^2 - 4\beta}| = 2\sqrt{\beta}$.

10.3 Consider the Schur decomposition $A = UTU^H$, where U is unitary and T is upper triangular. Since A is normal, we have $TT^H = T^H T$. One can verify that all off-diagonal terms of T vanish.

Appendix C
Further Readings

Machine learning has strong connections to statistics, optimization, numerical analysis, and mathematical analysis. In this book, we briefly discuss some selected topics and provide pointers for further study.

Interested readers are referred to more detailed studies on regression [5, 7], variance inflation factor [13], convex optimization [1, 4], LASSO [16], neural networks [3], deep learning [9], batch normalization [12], support vector machines [2, 18], the conjugate gradient method [10], the heavy ball method [17], singular value decomposition [6, 8], and principal component analysis [11, 14, 15].

The field of machine learning is rapidly evolving, and the references cited in the above paragraph are by no means comprehensive. Due to the limits of our knowledge, we only provide a selection of relevant references and apologize for omitting many other important works.

X.-S. Wang, C. Wang, *Machine Learning in Data Processing*, Forum for Interdisciplinary Mathematics, https://doi.org/10.1007/978-3-032-20855-2

References

1. S. Boyd and L. Vandenberghe, *Convex Optimization*. Cambridge University Press, Cambridge, UK, 2004.
2. C. Cortes and V. Vapnik, Support-vector networks, *Mach. Learn.* **20** (1995), 273–297.
3. G. Cybenko, Approximation by superpositions of a sigmoidal function, *Math. Control Signals Syst.* **2** (1989), 303–314.
4. I. Daubechies, M. Defrise, and C. De Mol, An iterative thresholding algorithm for linear inverse problems with a sparsity constraint, *Commun. Pure Appl. Math.* **57** (2004), 1413–1457.
5. N. R. Draper and H. Smith, *Applied Regression Analysis*, 3rd ed. Wiley, New York, 1998.
6. C. Eckart and G. Young, The approximation of one matrix by another of lower rank, *Psychometrika* **1** (1936), 211–218.
7. D. A. Freedman, *Statistical Models: Theory and Practice*. Cambridge University Press, Cambridge, UK, 2009.
8. G. H. Golub and C. F. Van Loan, *Matrix Computations*, 4th ed. Johns Hopkins University Press, Baltimore, MD, 2013.
9. I. Goodfellow, Y. Bengio, and A. Courville, *Deep Learning*. MIT Press, Cambridge, MA, 2016.
10. M. R. Hestenes and E. Stiefel, Methods of conjugate gradients for solving linear systems, *J. Res. Nat. Bur. Stand.* **49** (1952), 409–436.
11. H. Hotelling, Analysis of a complex of statistical variables into principal components, *J. Educ. Psychol.* **24** (1933), 417–441, 498–520.
12. S. Ioffe and C. Szegedy, Batch normalization: Accelerating deep network training by reducing internal covariate shift, in Proc. *32nd Int. Conf. Mach. Learn. (ICML)*, 2015, pp. 448–456.
13. G. James, D. Witten, T. Hastie, and R. Tibshirani, *An Introduction to Statistical Learning*, 2nd ed. Springer, New York, 2021.
14. I. T. Jolliffe, *Principal Component Analysis*, 2nd ed. Springer, New York, 2002.
15. I. T. Jolliffe and J. Cadima, Principal component analysis: A review and recent developments, *Phil. Trans. R. Soc. A* **374** (2016), 20150202.
16. R. Tibshirani, Regression shrinkage and selection via the lasso, *J. R. Stat. Soc. Ser. B* **58** (1996), 267–288.
17. B. T. Polyak, Some methods of speeding up the convergence of iteration methods, *USSR Comput. Math. Math. Phys.* **4** (1964), 1–17.
18. V. N. Vapnik, *Statistical Learning Theory*. Wiley, New York, 1998.

X.-S. Wang, C. Wang, *Machine Learning in Data Processing*, Forum for Interdisciplinary Mathematics, https://doi.org/10.1007/978-3-032-20855-2

Index

B
Backward propagation, 64, 71
Batch normalization, 69

C
Cauchy sequence, 103
Coefficient of determination, 24, 32
Conjugate gradient method, 83
Convex optimization, 34
Covariance, 19

D
Data
  dimension, 17
  nontrivial, 76
  separable, 76
  size, 17

E
Error, 19

F
Function
  activation, 60, 69
  antisymmetric, 3
  composite, 57
  continuous, 104
  convex, 34
  cost, 48
  differentiable, 104
  Fréchet differentiable, 12
  gradient, 12
  inner product, 83, 108
  likelihood, 19
  multilinear, 3
  normalized, 3
  sigmoid, 46

G
Gradient descent method, 49, 58
Gram-Schmidt process, 84, 93

H
Hyperplane, 76

I
Input, 17
  binary, 45
Iterative method, 39
  heavy ball method, 88
  Jacobi method, 87
  one-step, 86
  two-step, 88

K
Kronecker delta, 77

X.-S. Wang, C. Wang, *Machine Learning in Data Processing*, Forum for Interdisciplinary Mathematics, https://doi.org/10.1007/978-3-032-20855-2

L
Layer
  hidden, 60, 63
  input, 60, 63
  output, 60, 63
Least absolute shrinkage and selection operator (LASSO), 34
Least squares approximation, 20
Limit point, 103
Linear correlation, 25
Linear system, 27, 83, 86
Logic gate, 45

M
Matrix, 1
  characteristic polynomial, 108
  conjugate transpose, 93
  determinant, 3, 106
  eigenvalue, 108
  eigenvector, 108
  factorization, 94
  Hermitian, 106
  Hessian, 13
  identity, 2, 106
  invertible/nonsingular, 2, 106
  Jacobian, 13
  minor, 1
  multiplication, 106
  nonnegative definite/positive semidefinite, 108
  normal, 106
  orthogonal, 95
  positive definite, 83, 108
  principal component, 96
  product, 2
  rank, 9
  singular value, 95
  spectral radius, 108
  symmetric, 2, 106
  trace, 2, 107
  transpose, 2, 106
  unitary, 93, 106
Maximum likelihood estimation, 19
Mean, 19, 31

N
Neural network
  deep, 63
  depth, 63
  shallow, 60
Norm
  l^1, 34
  l^2, 99
  l^p, 108
  Frobenius, 99, 109
  matrix, 109
  vector, 75, 108
Normal distribution, 19, 31
  multivariate, 19

O
Objective function, 20
Observation, 18, 75
Optimization problem, 20, 48, 76
  constrained, 77
  minimax, 87, 88
Output, 17
  binary, 45
Overfitting, 32

P
Parameter, 18, 75
  feasible, 76
  learning rate, 39, 50
  normalized feasible, 76
  regularization, 41
  tuning/hyperparameter, 33
Prediction, 18, 46
Principal component analysis (PCA), 95
Projection, 21, 84, 96
Pythagorean theorem, 21, 97

R
Rayleigh quotient, 95
Regression
  linear, 17
  nonlinear, 48
  Ridge, 32
Regularization, 31
  l^1, 40
  l^2, 32, 40
Residual, 84

S
Sample variance, 24
Schur decomposition, 93
Set
  closed, 103
  compact, 104
  feasible, 75
  normalized feasible, 76

Singular value decomposition (SVD), 95
Steepest descent method, 39
Subderivative, 36
Subdifferential, 36
Sum of squared errors (SSE), 19
Support vector machine (SVM), 75

T
Taylor expansion, 13, 21, 33
Taylor's theorem, 14, 105

V
Variance, 19, 31
Variance inflation factor, 25, 32
Vectors, 1
  basis, 107
  linearly dependent, 11, 107
  linearly independent, 11, 107
  orthogonal basis, 107
  orthonormal, 96
  orthonormal basis, 99, 107
  standard basis, 2

MIX
Papier aus verantwortungsvollen Quellen
Paper from responsible sources
FSC® C105338

If you have any concerns about our products,
you can contact us on
ProductSafety@springernature.com

In case Publisher is established outside the EU,
the EU authorized representative is:
Springer Nature Customer Service Center GmbH
Europaplatz 3, 69115 Heidelberg, Germany

Printed by Libri Plureos GmbH
in Hamburg, Germany